IMAGINING THE UNIVERSE

IMAGINING THE UNIVERSE

A VISUAL JOURNEY

Edward Packard

A PERIGEE BOOK

Perigee Books are published by
The Berkley Publishing Group
200 Madison Avenue
New York, NY 10016

Published simultaneously in Canada

Library of Congress Cataloging-in-Publication Data
Packard, Edward, 1931–
    Imagining the universe: a visual journey / Edward Packard. —1st Perigee ed.
        p.   cm.
    "A Perigee book."
    ISBN 0-399-52124-0
    1. Outer space. 2. Solar system. 3. Space and time. 4. Cosmology. I. Title
QB500.P23 1994
500—dc20   94-17980
                                                            CIP

computer art and design: Wells Packard
book design and direction: Georgie Stout
designers: Douglas & Voss
production assistance: Richard Brightfield

cover concept: Edward Packard
cover design: David Bamford

Printed in the United States of America
10   9   8   7   6   5   4   3   2   1

Contents

INTRODUCTION

Of all the wonders of the universe, few are more awesome than its size. Numbers with long strings of zeros after them are of little help in comprehending it, nor are the terms scientists use to express the distances of outer space, such as "light-years" and "parsecs," and the distances of microscopic space, such as "micrometers" and "nanometers."

Similarly, the vast stretches of time from the beginning of the universe to the present mean little when expressed in "eons" or "eras" or, for that matter, in millions or billions of years.

This book approaches space and time visually rather than mathematically. In Parts I and II the distances and sizes of objects in outer space and in microscopic space are compared with the sizes of familiar objects ranging from a grain of sand to the Earth itself. In Part III, time is seen as an object moving across the surface of the Earth. We follow it — and the course of events — as it moves one mile every million years.

This book, then, is a visual trip through space and time — one that we hope will enable you, in a way, to "imagine the universe."

PART I

LOOKING OUTWARD

THE SOLAR SYSTEM

The Earth's immediate neighbors are the sun, eight other planets and their satellites, and countless other objects that revolve around the sun.

The solar system is so vast that to imagine it we must shrink the universe drastically, so much that the Earth will only be the size of a baseball park.

ASSUME THE EARTH IS THE SIZE OF A BASEBALL PARK

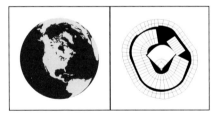

The shrunken Earth nestled in Candlestick Park, looking east over San Francisco Bay.

THE EARTH IS THE SIZE OF A BASEBALL PARK

With the Earth nestled in Candlestick Park, the moon is almost on the other side of San Francisco Bay, near the Oakland International Airport.

Of course, orbiting bodies such as the moon and the planets may be found anywhere along their orbits. Thus, the moon in this illustration might just as well be hovering over the San Francisco–Oakland Bay Bridge.

The moon may have formed from debris kicked up by the gigantic impact of a planetoid upon the Earth soon after the Earth itself had formed. Gradually the Earth's gravity slowed the moon's rotation to a stop. Since then, Earth-bound viewers have been able to see only one side of the moon.

From the moon "full Earth" presents a spectacularly beautiful sight, appearing as a disc four times as big across as the moon appears to us. Earth's continents, oceans, and clouds are plainly visible.

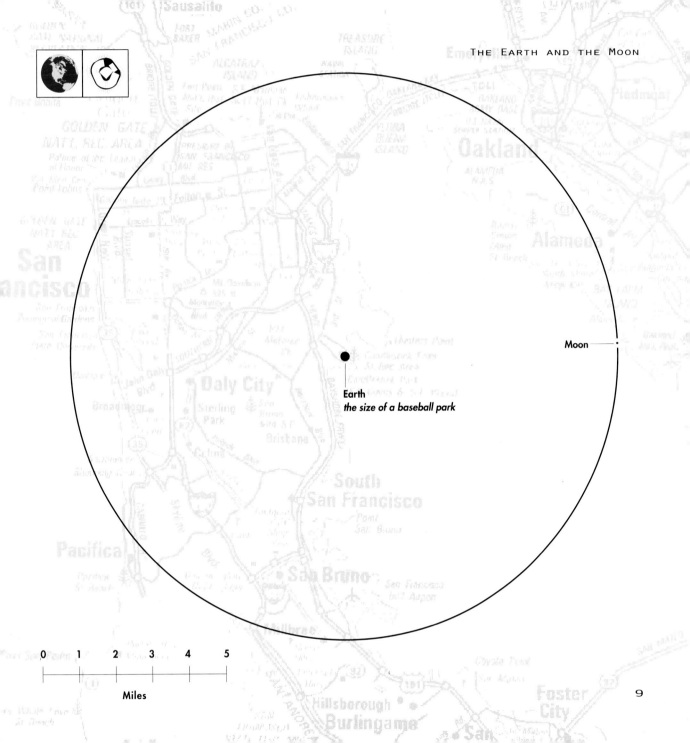

Moon

Earth
the size of a baseball park

0 1 2 3 4 5

Miles

With the Earth still nestled in Candlestick Park, we find Venus, the size of a slightly smaller park, 875 miles away near El Paso, Texas.

Mars, about half the size of a baseball park, is floating over the Gulf of Alaska.

None of these planets is big enough to be shown even as a dot on the map. The arrows merely point to their locations.

Venus is perpetually obscured by dense clouds. Its surface remained a mystery until space probes relayed radar images revealing mountains, volcanoes, and rolling uplands. Where oceans might be are only "lowlands." Unlike Earth, Venus is geologically inactive.

Venus is an extreme case of a "greenhouse disaster." Heat from the sun is trapped under an extremely dense atmosphere, one rich in carbon dioxide but lacking oxygen. The planet's high temperature and surface air pressure would pose a seemingly insurmountable obstacle to visiting astronauts.

Mars, with a diameter of little over half that of the Earth's, is a dry, cold planet with an extremely thin atmosphere. Ice caps of frozen carbon dioxide cover the poles, advancing and receding with the Martian seasons. From Mars, Earth would look, much as does Venus to us, like a very brilliant star in the evening or predawn sky.

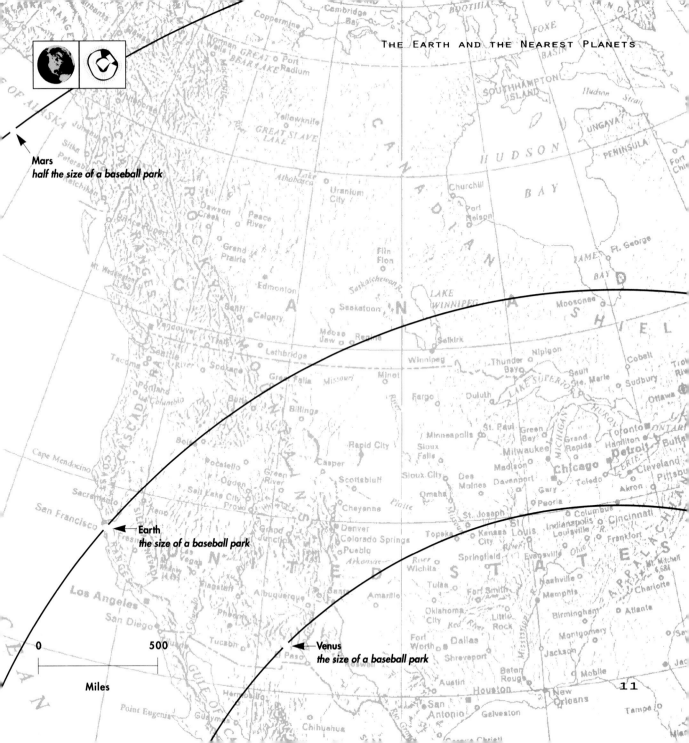

Mars
half the size of a baseball park

Earth
the size of a baseball park

Venus
the size of a baseball park

0   500

Miles

To find all the inner planets and the sun we must look farther afield. Mercury, about a third the size of our ball-park-sized Earth, is in central Mexico. The sun, twenty-seven miles in diameter, is off the coast of Costa Rica.

The sun formed about five billion years ago out of an interstellar cloud of gas and dust that collapsed with increasing speed under the force of gravity. When density and pressure reached a critical mass, thermonuclear reactions began, and a star was born.

The sun has maintained a relatively stable size for billions of years because the explosive force of nuclear reactions within it has been balanced by the force of gravity. Stars larger than the sun burn out a great deal more rapidly, the largest in a few million years.

Mercury is a small, hot, moonlike planet. Because it's so close to the sun, it's never visible except just after sunset or just before sunrise.

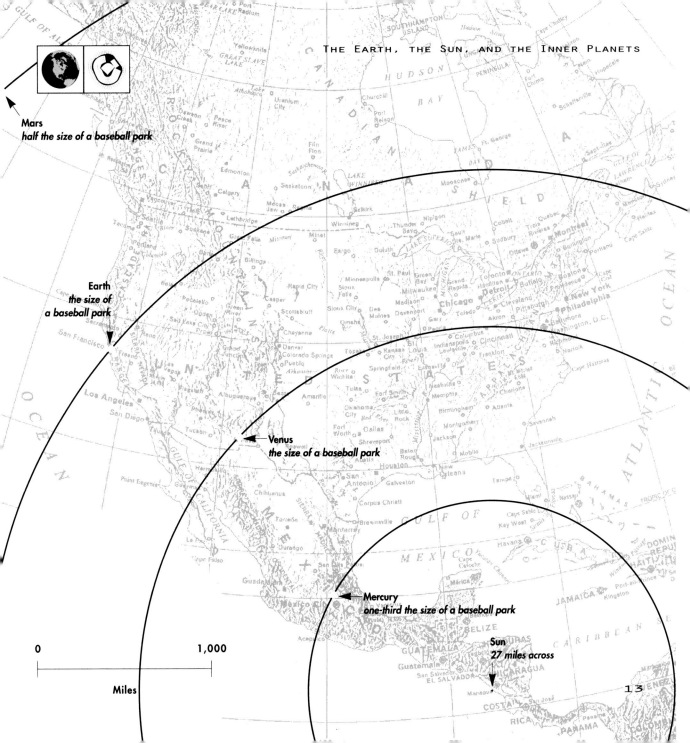

Mars
half the size of a baseball park

Earth
*the size of
a baseball park*

Venus
the size of a baseball park

Mercury
one-third the size of a baseball park

Sun
27 miles across

0   1,000

Miles

13

The Earth, still nestled in Candlestick Park, with the sun and inner planets clustered nearby.

Both Jupiter and Saturn are well out in space.

← Saturn
2.4 miles across

The planets shown here were identified as distinct from stars in ancient times because of their eccentric movements and varying luminosity. The stars, by contrast, seemed "fixed" in the sky. The great achievement of Copernicus, 500 years ago, was in showing that the planets, including Earth, revolved around the sun.

Jupiter and Saturn are composed largely of hydrogen and helium and are vastly larger than the Earth. They both have numerous moons, some of which are geologically active. Jupiter's largest moons were discovered by Galileo in 1610 with the aid of a primitive telescope. Four of them, lined up in the same plane, are easily visible with binoculars.

Jupiter
2.8 miles across

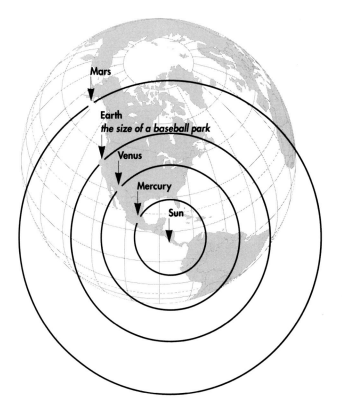

Mars

Earth
the size of a baseball park

Venus

Mercury

Sun

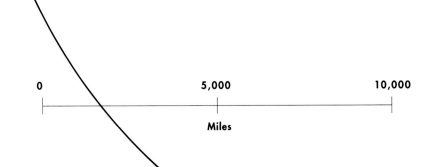

0   5,000   10,000

Miles

Thus far, we have shrunk the universe so much that the Earth is only the size of a baseball park. To imagine the entire solar system, we must shrink the Earth a great deal more, so much that our planet is only the size of a baseball.

ASSUME THE EARTH IS THE SIZE OF A BASEBALL

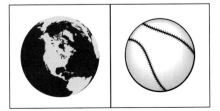

The baseball-sized Earth floating over home plate in Candlestick Park. The four base paths (the diamond) are each ninety feet long. The distances shown on the left, center, and right extremities of the outfield are measured from home plate.

The moon, the size of a cherry, is about seven and a half feet away from Earth. The sun and even the nearest planets are outside the ball park.

THE EARTH IS THE SIZE OF A BASEBALL

400'

335'

335'

90'

Moon
the size of a cherry

◄ Earth
the size of a baseball

19

Planets do not orbit the sun in a perfect circle, as shown here, but in an ellipse. The Earth's orbit is almost circular, varying in distance from the sun by only 2 percent a year. By contrast, Mars's distance from the sun varies up to 20 percent.

The closer to the sun a planet is, the faster it must move to offset the sun's gravity. A year on Venus lasts less than eight months, a Martian year twenty-two.

Venus (also baseball sized), shown here farther along in its orbit, is now about a half mile from Earth. The sun (twenty-seven feet across) is almost three-quarters of a mile from Earth. Golf ball–sized Mars, shown near its extreme distance from Earth, is almost a mile and a half from Earth. At its closest approach, it would be a third of a mile away.

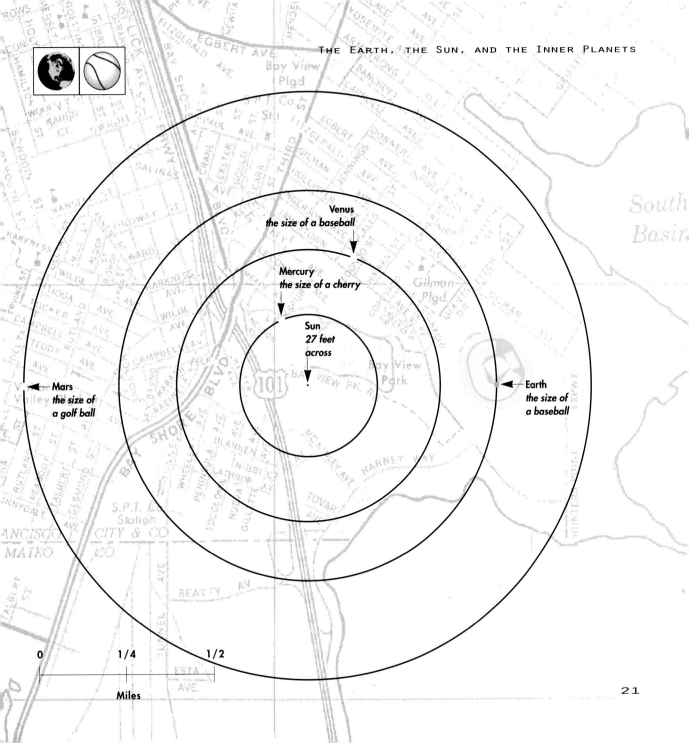

Venus
the size of a baseball

Mercury
the size of a cherry

Sun
27 feet across

Mars
*the size of
a golf ball*

Earth
*the size of
a baseball*

0   1/4   1/2

Miles

With the Earth only the size of a baseball (floating over home plate in Candlestick Park), the visible planets and the sun can still be found in the city of San Francisco.

If the planets were at the locations in their orbits shown in the illustration, Earth viewers would be unable to see Jupiter or Saturn because they would be behind, or nearly behind, the sun. Mars would be visible from late evening until dawn.

Saturn's famous rings are not solid, though they look so when viewed through a telescope, but are composed of countless tiny particles, mostly of dust and ice.

THE EARTH IS THE SIZE OF A BASEBALL

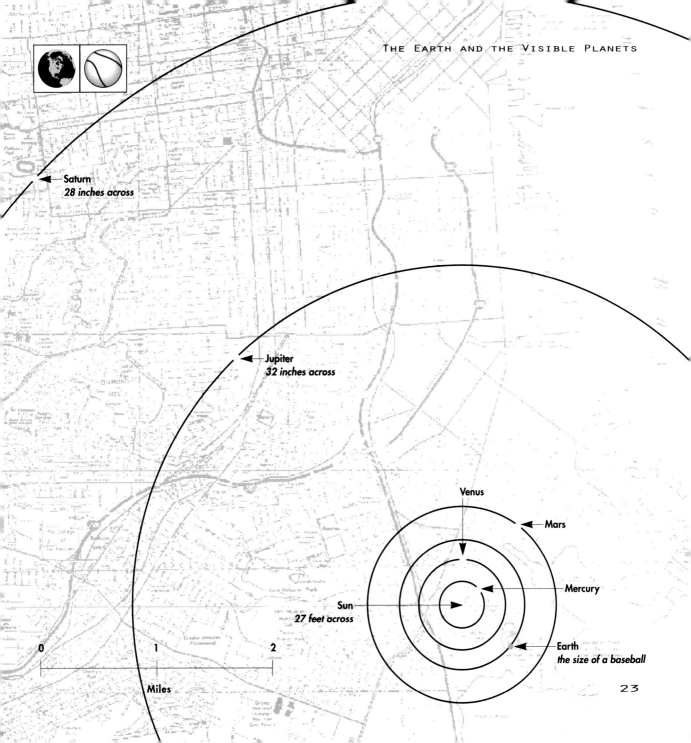

Saturn
28 inches across

Jupiter
32 inches across

Venus

Mars

Mercury

Sun
27 feet across

Earth
the size of a baseball

0   1   2

Miles

23

◄— Pluto

◄— Neptune

With the Earth the size of a baseball (floating over home plate in Candlestick Park), the solar system occupies the entire Bay area, and then some.

Shown here are the principal components of the solar system, the sun and nine planets. Not shown are moons of the various planets, rings, comets, particles of dust and other material, and asteroids ranging up to 500 miles across.

In the largest sense the solar system encompasses not only the planets, but also the heliosphere, the vast region that is affected by the "solar wind" — the gas emanating from the sun in all directions — and by a tremendous number of comets, most of which have highly elliptical orbits ranging far beyond the outer planets. Here we follow tradition in treating the solar system as being bound by the orbit of Pluto, the outermost planet of the sun.

THE EARTH IS THE SIZE OF A BASEBALL

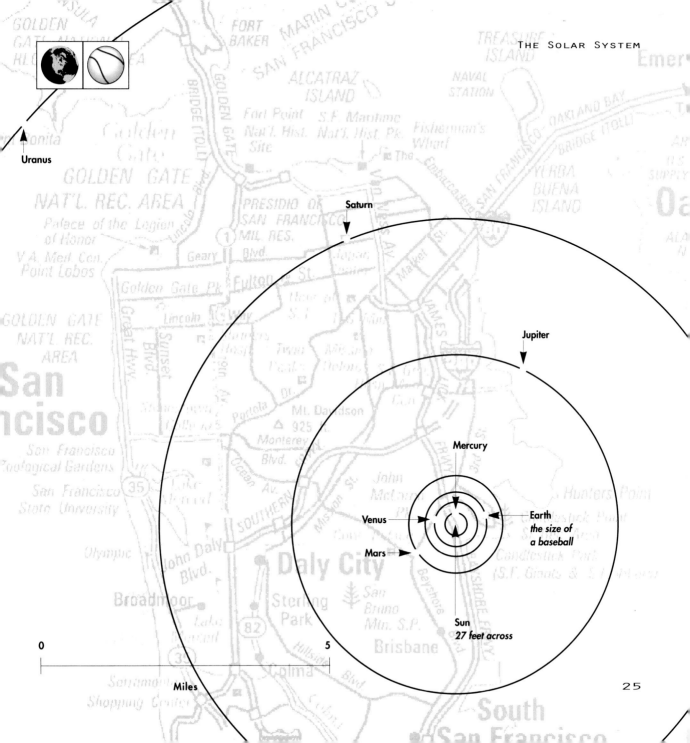

Uranus

Saturn

Jupiter

Mercury

Venus

Mars

Earth
*the size of
a baseball*

Sun
27 feet across

0   5

Miles

25

OUR STELLAR COMMUNITY

With the Earth shrunk to the size of a baseball (floating over home plate in Candlestick Park), Pluto, the farthest planet (now the size of a golf ball), is thirty miles out over the Pacific Ocean. One might imagine that the nearest stars would be much more distant than that — somewhere in Utah, perhaps, or in Mexico. But the nearest star would not be on a map of the world, or close to it.

To begin to imagine the size of our stellar community we must shrink the universe a great deal more, so much that Earth is only the size of a fine grain of sand.

ASSUME THE EARTH IS THE SIZE OF A GRAIN OF SAND

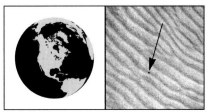

Because of its huge mass, the sun captured nine planets in orbit during the process of its formation. We can expect that similar stars have arrays of planets orbiting them as well. Many of these planets have (or have had, or will have) conditions favorable for the appearance and development of life, unless, as some think, those conditions are so precise as to be unique.

Looking down on Candlestick Park, we find the grain-of-sand-sized Earth floating over home plate.

Venus is another grain of sand, in the catcher's mitt. The sun, the size of a golf ball, is ten feet behind home plate. Mars is a speck of dust a few feet from the plate, Jupiter, the size of an apple seed, three quarters of the way to the mound, and Saturn, a smaller apple seed, near first base. Uranus and Neptune, each the size of pinheads, are respectively in shallow left field and in left center field. Pluto is a speck of dust near the center-field wall.

The baseball field–sized solar system is almost empty. There is nothing much in it but a golf ball–sized sun and nine planets no larger than apple seeds, pinheads, sand grains, and dust specks.

THE EARTH IS THE SIZE OF A GRAIN OF SAND

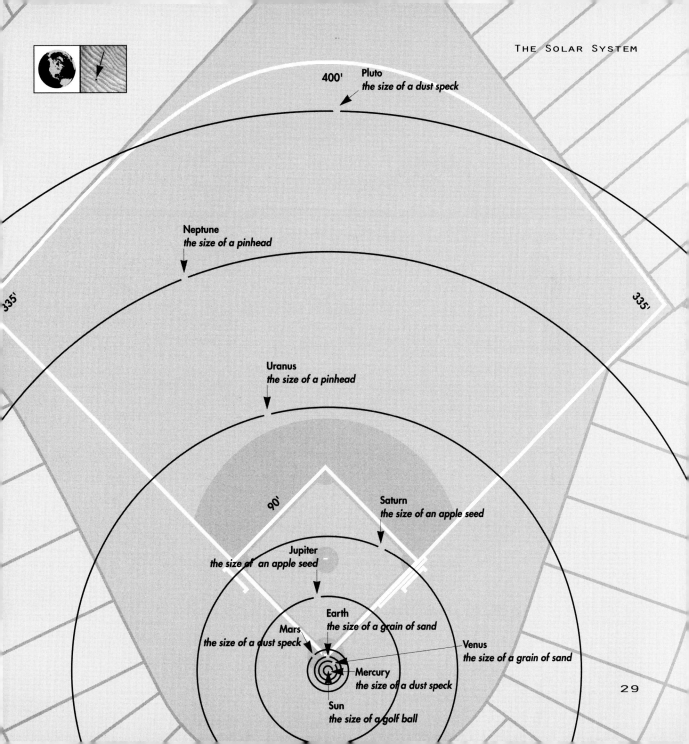

Pluto
the size of a dust speck

400'

Neptune
the size of a pinhead

335'

335'

Uranus
the size of a pinhead

90'

Saturn
the size of an apple seed

Jupiter
the size of an apple seed

Earth
the size of a grain of sand

Mars
the size of a dust speck

Venus
the size of a grain of sand

Mercury
the size of a dust speck

Sun
the size of a golf ball

29

We have reduced the Earth to the size of a grain of sand and thereby fit the entire solar system into the playing field at Candlestick Park. If we set out from our planet (floating over home plate) and cruise though the infield past the dust speck that is Mars and the apple seeds that are Jupiter and Saturn, through the outfield past the pinhead-sized Uranus and Neptune, then past the speck of dust that is Pluto, and continue on through space to the nearest star, we will not find it until we reach the Grand Canyon.

Unlike the sun, most stars are members of star systems with two or more stars held in orbits around each other by gravitational attraction. Proxima Centauri, the nearest star, is associated in this fashion with a pair of much larger stars, Alpha Centauri A and B, which are so close together that they appear to the naked eye as a single bright star.

One component of Alpha Centauri is slightly larger than the sun, the other slightly smaller than the sun. Proxima Centauri, the third component, is much smaller than either. Although it is this star that is closest to us, it is so dim that it cannot be seen with the naked eye. By chance, our nearest neighbor is almost as small as a star can be.

Earth
the size of a grain of sand

Sun
the size of a golf ball

Alpha Centauri A
the size of a golf ball

Proxima Centauri →
the size of a peppercorn

Alpha Centauri B
the size of a Ping-Pong ball

0   250   500

Miles

31

There is no natural boundary to our "stellar neighborhood." It is shown here as the region containing approximately the twenty-five nearest stars. Despite the proximity of these stars, fewer than half of them are visible to the naked eye.

Stars range tremendously in size and brightness. Most of them are relatively dim compared with the sun. But the ones visible in the night sky are with few exceptions bigger and more brilliant than the sun.

Within about fifteen hundred miles of the grain of sand that is the Earth, we'll find about a dozen stars. Most of these nearby stars are smaller and dimmer than our golf ball–sized sun. But two of them are larger and brighter than the sun. One of these, Sirius, is as big as a baseball. It is the brightest star in Earth's sky. Another, somewhat more distant one, Procyon, is even larger.

Stars and galaxies in this book are shown here as if they were all on the same plane. In reality, of course, they are to be found in all directions from Earth. And these illustrations show relative distances, not actual directions, of stars and galaxies from us.

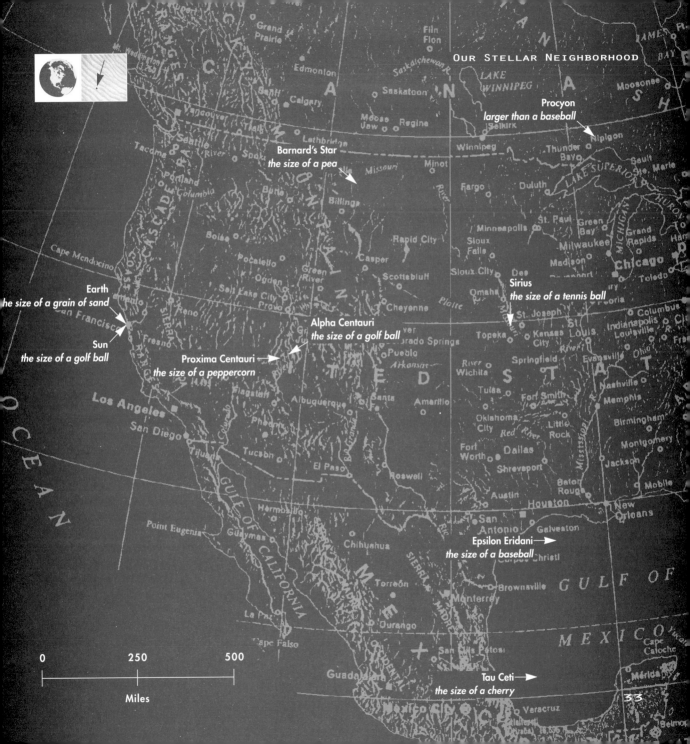

Procyon
larger than a baseball

Barnard's Star
the size of a pea

Sirius
the size of a tennis ball

Earth
the size of a grain of sand

Sun
the size of a golf ball

Alpha Centauri
the size of a golf ball

Proxima Centauri
the size of a peppercorn

Epsilon Eridani
the size of a baseball

Tau Ceti
the size of a cherry

0   250   500

Miles

33

With the Earth the size of a grain of sand and the sun the size of a golf ball (both of them hovering over home plate in Candlestick Park), thousands of stars surround us, occupying a volume far larger than the globe. They range from the size of a peppercorn to the size of a basketball. Only a few of the prominent ones are indicated here.

Our "stellar community" is an arbitrary term. It is shown here as the region containing most of the brightest stars in the night sky.

Interstellar space is not completely empty. It contains clouds of hydrogen, helium, and other elements. These clouds are in no way like clouds we're familiar with. They are far thinner than the atmosphere even hundreds of miles above the Earth.

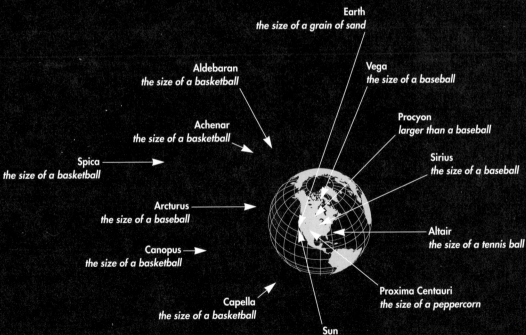

Earth
the size of a grain of sand

Aldebaran
the size of a basketball

Vega
the size of a baseball

Achenar
the size of a basketball

Procyon
larger than a baseball

Spica
the size of a basketball

Sirius
the size of a baseball

Arcturus
the size of a baseball

Altair
the size of a tennis ball

Canopus
the size of a basketball

Proxima Centauri
the size of a peppercorn

Capella
the size of a basketball

Sun
the size of a golf ball

0   10,000

Miles

OUR GALAXY

By shrinking the universe until the Earth is only the size of a grain of sand, we have endeavored to imagine first the solar system and then our stellar community. If we are to imagine our entire galaxy — the Milky Way — we must shrink the universe a great deal more, so much that the sun is only the size of a grain of sand.

ASSUME THE SUN IS THE SIZE OF A GRAIN OF SAND

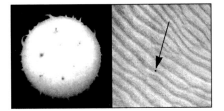

To appreciate what it would mean to shrink the sun to the size of
a grain of sand, we must first have a sense of how large the sun is.

The sun looks no larger than the moon in the sky, yet its
diameter is more than a hundred times that of the Earth's.

Earth

In the fall of 1977, taking advantage of favorable alignment of the planets, NASA launched two space probes on a mission to explore the solar system. *Voyager 2*, the most successful of them, passed close to Jupiter in July 1979 and Saturn in August 1981. Despite radio and mechanical problems, engineers were able to rehabilitate and even improve the probe. It passed close to Uranus in 1986 and Neptune in 1989, where it sent back stunning pictures of a planet that had been no more than a fuzzy disc in even the largest telescopes.

The sun is the size of a grain of sand (floating over home plate in Candlestick Park). The Earth, shrunk so much that it's microscopic, is an inch away. Nine planets — the entire solar system — whirl in a six-foot circle around the grain-of-sand-sized sun.

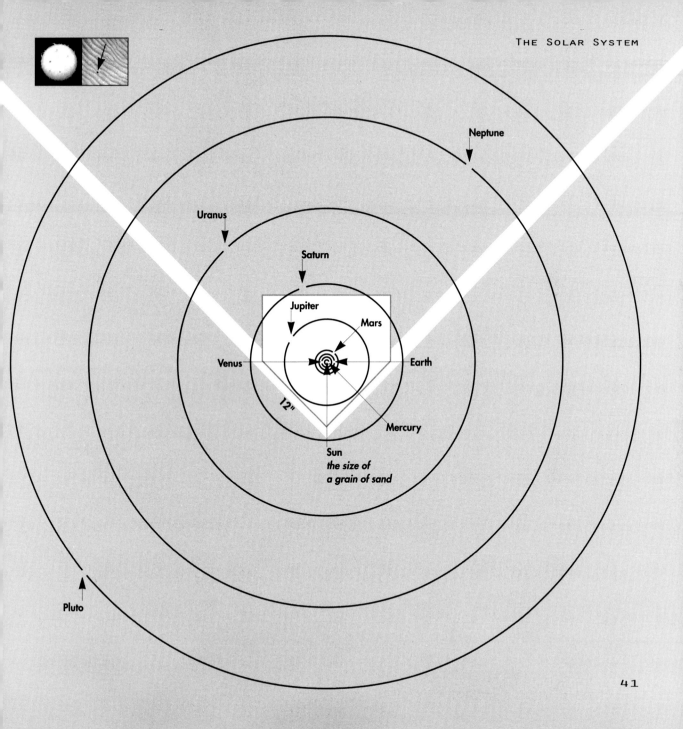

Neptune

Uranus

Saturn

Jupiter

Mars

Venus

Earth

12"

Mercury

Sun
*the size of
a grain of sand*

Pluto

With the grain-of-sand-sized sun floating over home plate in Candlestick Park (and microscopic Earth an inch from it), the nearest star (even smaller than a grain of sand) is about five miles away.

On this scale, our stellar community extends a couple of hundred miles or so from San Francisco.

Proxima Centauri, earlier noted for its dimness, is representative of a huge class of stars known as red dwarfs. The cooler, dimmer stars have a dull red color like glowing embers. Hotter stars are yellow like the sun. Still hotter and brighter stars are blue, the hottest tending toward white.

Dimmer and smaller even than red dwarfs are the brown dwarfs, stars large enough to generate heat, but too small to ignite — to initiate thermonuclear reactions. Astronomers have not been able to view a brown dwarf even with the largest telescopes, yet these marginal stars may be the most numerous of all.

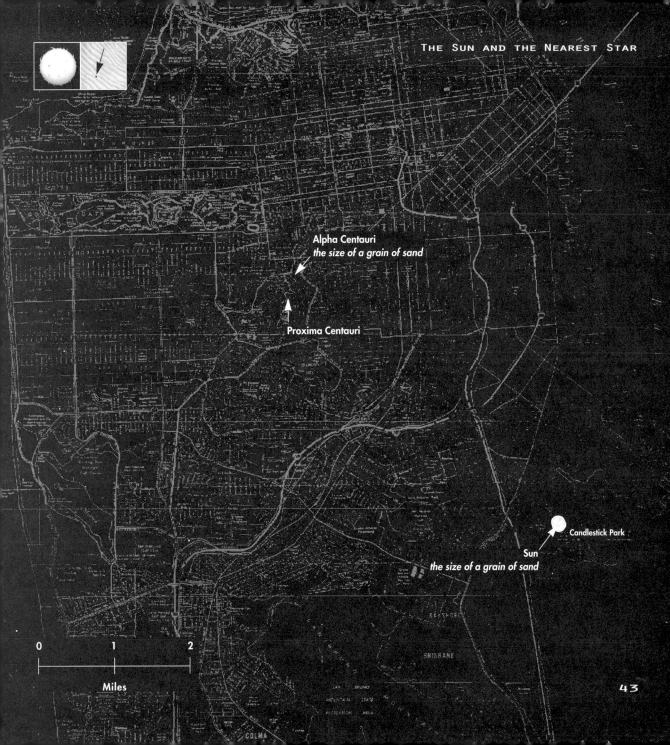

Alpha Centauri
the size of a grain of sand

Proxima Centauri

Candlestick Park

Sun
the size of a grain of sand

0   1   2

Miles

43

The grain-of-sand-sized sun is still floating over home plate in Candlestick Park. By expanding our map to show the whole globe and nearby space, we find hundreds of millions of stars. Some are pea-sized or cherry-sized giants, but most of them, like the sun, are no larger than a grain of sand.

Large as our map is, and as much as we've shrunk the universe, we have so far imagined only our local area, a tiny fraction of our galaxy.

The area in the vicinity of the solar system is relatively thinly populated with stars. Observers on planets of most stars would have a far more dazzling sight. Others, in the middle of dense star clusters, might not experience the night sky at all — they would be surrounded by suns!

Most stars are billions of years old, but the most massive stars last only a few million years before their existence is ended by an explosion so violent that for a few days the star — called a supernova — will outshine the entire galaxy. Such an event happening in our galaxy is likely to be seen from Earth only once every several hundred years. If a star in our stellar neighborhood became a supernova (an unlikely occurrence) the radiation burst from it might be sufficient to extinguish all surface life on Earth.

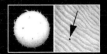

Sun
the size of
a grain of sand

This photograph gives some idea of how the Milky Way galaxy would look to some very distant alien. Since our galaxy could only be photographed by someone outside it, the image shown is of a galaxy somewhat similar to ours that lies in the direction of, and far beyond, the Big Dipper.

Our galaxy, a large one with perhaps as many as 300 billion stars, is more or less a spiral disc, thickening at the core. From our vantage near the inside of a spiral arm, we can look toward the star-packed center. On a clear dark night we can see the galactic plane, which like the galaxy itself is called the Milky Way. Binoculars or a small telescope gives some idea of the great density of stars as we approach the galactic center, although dark spots in the field of view, caused by dust clouds, block off the light of countless more stars.

To imagine our entire galaxy — the Milky Way — remember that the sun has been reduced to a grain of sand (floating over home plate in Candlestick Park). As shown on page 43, the nearest star is another grain of sand about five miles away. As shown on page 45, hundreds of millions of stars, most of them no larger than a grain of sand, lie within a few thousand miles of San Francisco.

In the illustration on the right we pull back from the globe far enough to see the entire galaxy. In looking at the globe bear in mind that it shows the sun shrunk not just to the size of a grain of sand dropped on the page, but to the size of a grain of sand in San Francisco. Not only is the grain of sand too small to be shown as a dot, the entire city of San Francisco is too small to be shown as a dot.

Sun
the size of
a grain of sand

The Universe

We have previously shrunk the universe so much that the sun is only the size of a grain of sand. Yet even on this scale our galaxy stretches far beyond the globe. Our galaxy, however, occupies only an infinitesimal part of the universe. If we are to imagine the entire universe, we will have to shrink it still more, so much that the solar system is only the size of a grain of sand.

ASSUME THE SOLAR SYSTEM IS THE SIZE OF A GRAIN OF SAND

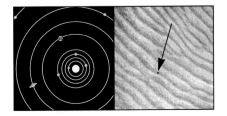

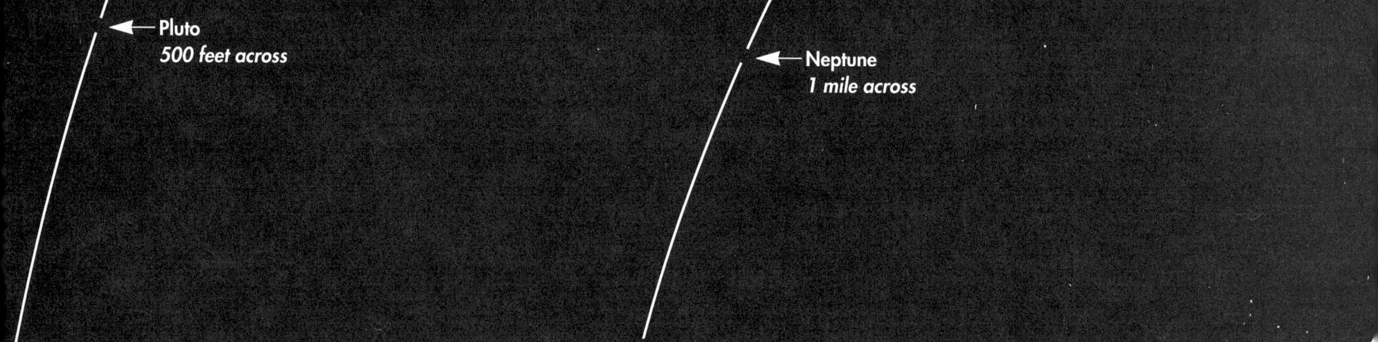

← Pluto
500 feet across

← Neptune
1 mile across

Before shrinking the solar system to the size of a grain of sand, envision again how vast the solar system is. For that purpose, we'll first restore the Earth to the size of a baseball park.

With the Earth the size of a baseball park in San Francisco the solar system extends far beyond the globe.

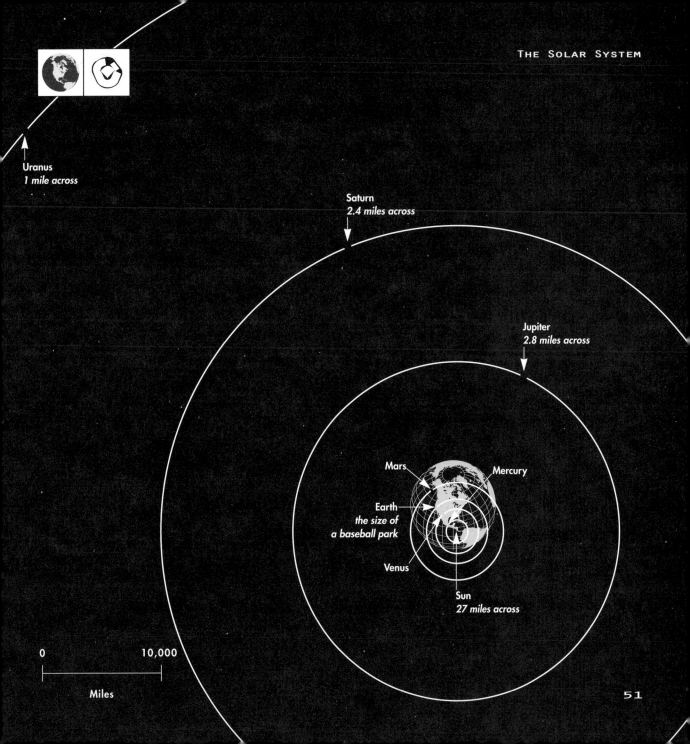

Uranus
1 mile across

Saturn
2.4 miles across

Jupiter
2.8 miles across

Mars

Mercury

Earth
*the size of
a baseball park*

Venus

Sun
27 miles across

0   10,000

Miles

We temporarily restored the Earth to the size of a baseball park in order to recall how vast the solar system is.

Imagine then, the entire solar system shrunk to the size of a grain of sand.

With the universe so drastically shrunk that the solar system is only the size of a grain of sand in Candlestick Park, the nearest star is only three feet away and our stellar community could fit in the infield.

Orbiting the millions of stars in our stellar community are even greater millions of planets. How many of them harbor life? Some years ago the United States funded a serious effort to detect signals giving evidence of intelligent beings in outer space. Nothing was found, and funds for the project were withdrawn. Private sources subsequently financed the search on a scaled-back basis.

Unlikely as life may be on any one planet, space is so vast — there are so many stars, so many planets — that it would be even more strange if life on Earth were unique. But even if there is life on other planets or elsewhere, it seems unlikely that current methods of searching for it will succeed. The stars are so widely dispersed that even with heroic efforts we could survey only the relatively few stars close by that most resemble the sun. There seems only a minuscule chance that creatures on one of their planets are beaming sufficiently powerful signals to be detected on Earth.

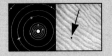

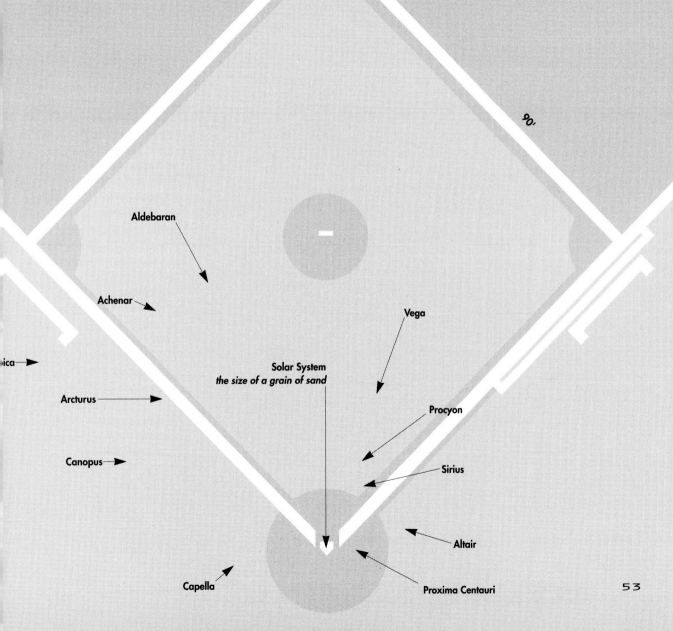

90'

Aldebaran

Achenar

Vega

ica

Arcturus

Solar System
the size of a grain of sand

Procyon

Canopus

Sirius

Altair

Capella

Proxima Centauri

With the grain-of-sand-sized solar system floating over home plate in Candlestick Park, the nearest star is three feet away. Virtually all the stars we can see in the night sky fit in the park, and the hundreds of millions of stars in the local area of our galaxy fit in a small corner of San Francisco.

The sun and all the other stars in the galaxy revolve around the galactic core, a journey that for the sun lasts two hundred and forty million years. The stars are also moving relative to each other. They may be thought of like fish swimming in different directions while being carried along in the same current, or more precisely, in the same eddy in the main current, the motion of the galaxy through space.

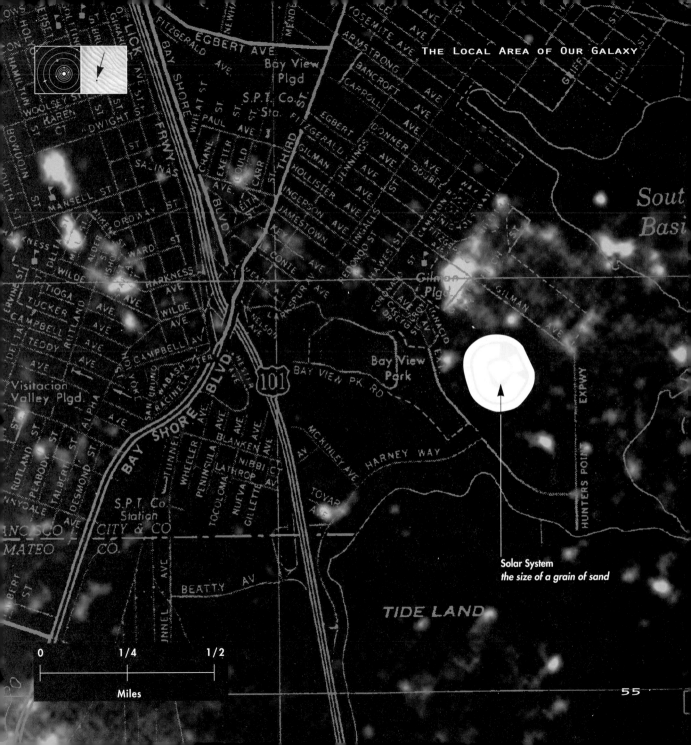

Sout
Basi

Solar System
the size of a grain of sand

0   1/4   1/2

Miles

TIDE LAND

With the solar system the size of a grain of sand (floating over home plate in Candlestick Park), our galaxy spans San Francisco Bay.

Looking up at the sky on a clear, moonless night we see an awesome array of stars. But this experience can give only the feeblest notion of the scope of our galaxy. For every star we can see, there are a hundred million we cannot see.

If we were to approach the center of the Milky Way, we would encounter increasingly thick aggregations of stars. At the heart of the galaxy may lie a black hole, a concentration of matter so dense that light itself is unable to escape its gravity.

Eventually the galaxy will exhaust its supply of raw material, the interstellar clouds of gas and dust that provide the stuff from which stars are made. Until then, new stars, new planets, and perhaps new forms of life will be constantly in the making.

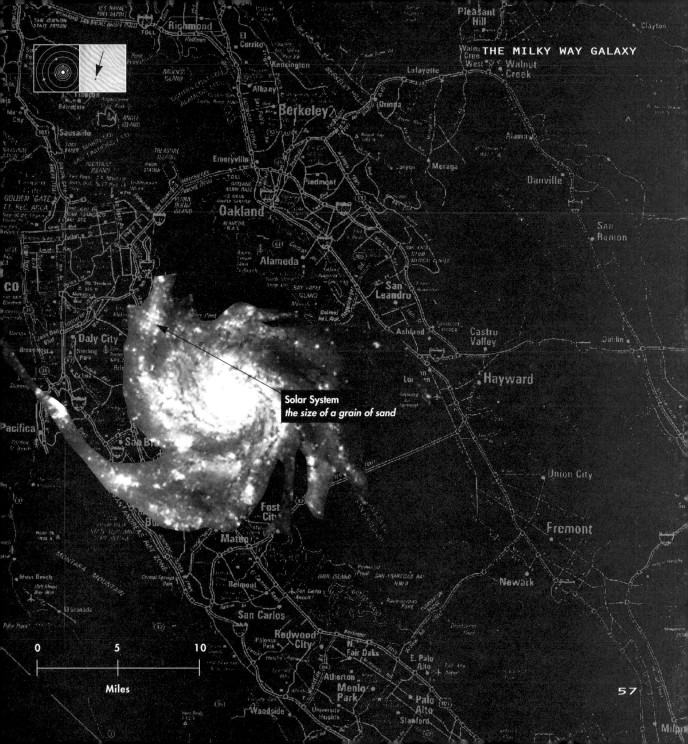

Solar System
the size of a grain of sand

0   5   10

Miles

With the solar system the size of a grain of sand, our galaxy spans San Francisco Bay. Our neighboring large galaxy — Andromeda — is three hundred miles away in the middle of Nevada.

The Andromeda galaxy is the farthest object that can be seen with the naked eye. It is visible from north latitudes much of the year, appearing as a fuzzy patch of light. Andromeda is somewhat larger than our own large galaxy and, like it, a spiral. Its arms are not readily apparent because we see it from an oblique angle.

The galaxies shown in this illustration are in what astronomers call the "local group."

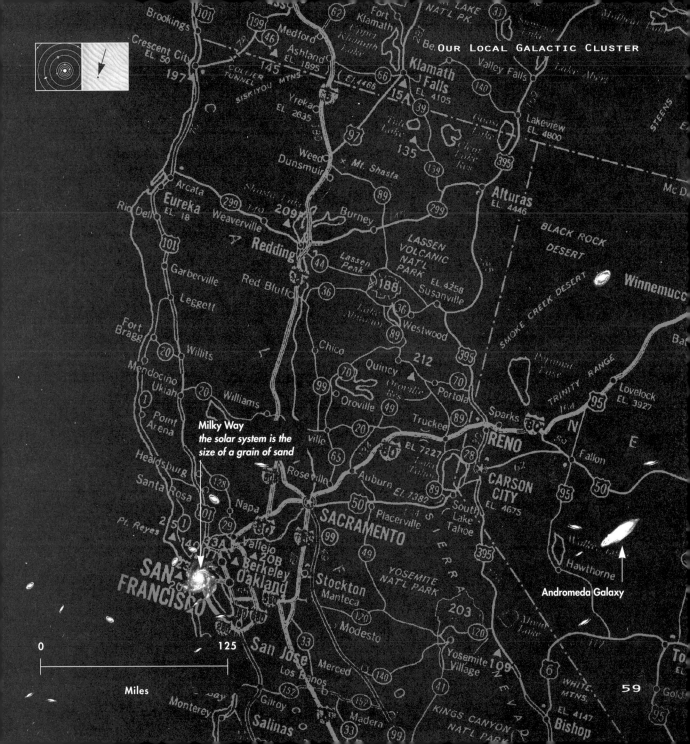

Milky Way
*the solar system is the
size of a grain of sand*

Andromeda Galaxy

0   125

Miles

With the solar system the size of a grain of sand in San Francisco, the Milky Way spans San Francisco Bay. Our local galactic cluster is several hundred miles across. Our local supercluster extends far beyond the limits of the globe.

Our local cluster of galaxies is itself part of a vastly larger aggregation of galaxies, called the Virgo supercluster because its center lies in the direction of the constellation Virgo, or, to put it more accurately, the constellation Virgo lies in the direction of the center of the supercluster.

The large-scale structure of the universe is the subject of increasing study. It has become evident that galaxies are not uniformly distributed in space but are arrayed, though irregularly, somewhat like particles on the surface of bubbles, leaving hollows relatively devoid of matter.

According to generally accepted theory, the galaxies are moving progressively farther away from each other, in some sense like polka dots on an expanding balloon. Notwithstanding this grand progression, galaxies in clusters and superclusters tend to remain in one another's company because of mutual gravitational attraction.

THE SOLAR SYSTEM IS THE SIZE OF A GRAIN OF SAND

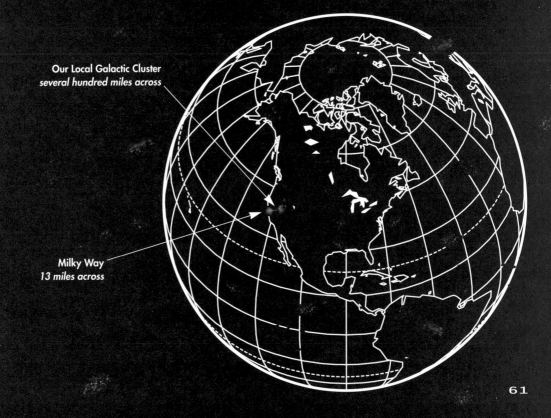

Our Local Galactic Cluster
several hundred miles across

Milky Way
13 miles across

61

Most scientists believe that the universe is between ten and twenty billion years old, that it originated from a single point — "the big bang" — and that it has been expanding ever since. As yet there is no consensus as to how much it has expanded — how large it has become; we have only "models" of the universe believed to be consistent with observations. As new theories are developed and new observations made, one such model, or some new model, may become accepted as describing the universe's origin, size, and eventual fate.

To imagine the universe, first imagine where San Francisco is on the tiny globe below. Then imagine that the entire solar system is a grain of sand floating over home plate in Candlestick Park.

Our galaxy, the Milky Way, consisting of hundreds of billions of stars, covers part of San Francisco (including Candlestick Park) and extends across San Francisco Bay. It is not big enough to show as a dot on the globe. Each dot in the pages that follow represents not a star, or even a galaxy, but a cluster of galaxies.

Hundreds of billions of galaxies lie in all directions from us for a distance as far as, if not farther than, shown here.

Milky Way ———→ ⬤ Earth
13 miles across

Solar System
*the size of a grain of
sand in San Francisco*

Superclusters of Galaxies

Space-time is "curved," and for this reason, at the end of our journey across the universe, we may discover that we have returned to our microscopic Earth, deep inside the grain-of-sand-sized solar system floating over home plate in San Francisco.

Although the universe may be finite, it is unbounded. No matter where one is, one appears to be at its center. We cannot talk usefully about what lies beyond the limits of the universe, because space itself can only be defined by the matter it contains.

As to what matter space does contain there is a great mystery. Most of the matter in space — perhaps 90 percent of it — is invisible, in fact undetectable except by observing the effect of its gravitational influence. The nature and origin of this "dark matter" — the universe's "missing mass" — is a central question in the search to understand the cosmos.

Milky Way ⟶ Earth
13 miles across

Solar System
*the size of a grain of
sand in San Francisco*

PART II

LOOKING INWARD

THE REALM OF CELLS

Looking outward, we shrank the Earth, the sun, and finally the entire solar system to the size of a grain of sand so that we could visualize the relative distances from us of planets, stars, and galaxies.

Looking inward we shall reverse the process so that we can compare the sizes of microscopic objects to the sizes of familiar ones.

We begin our journey into the microscopic world by expanding a baseball, so much that it is the size of a baseball park.

ASSUME A BASEBALL IS THE SIZE OF A BASEBALL PARK

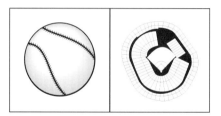

A baseball nestled in Candlestick Park

A BASEBALL IS THE SIZE OF A BASEBALL PARK

75

Biting midges, gnats, no-see-ums — call them what you will — can slip through screens that stop mosquitoes. They are visible to the naked eye under good conditions, which are rarely present when they bite.

By expanding a baseball to the size of a baseball park, a vantage over the playing field brings objects as close as the naked eye can see. The baseball dwarfs the playing field. An ant dominates the outfield. A grain of sand, now four feet across, is plainly visible, as are pond creatures that we could normally see only close up with sharp eyes and in a strong light.

Water bears are common in ponds, ditches, and moisture-laden plants. They have tubular mouth parts through which they suck juices of plants and tiny animals, which they pursue on their eight stubby legs.

Grasshopper

A BASEBALL IS THE SIZE OF A BASEBALL PARK

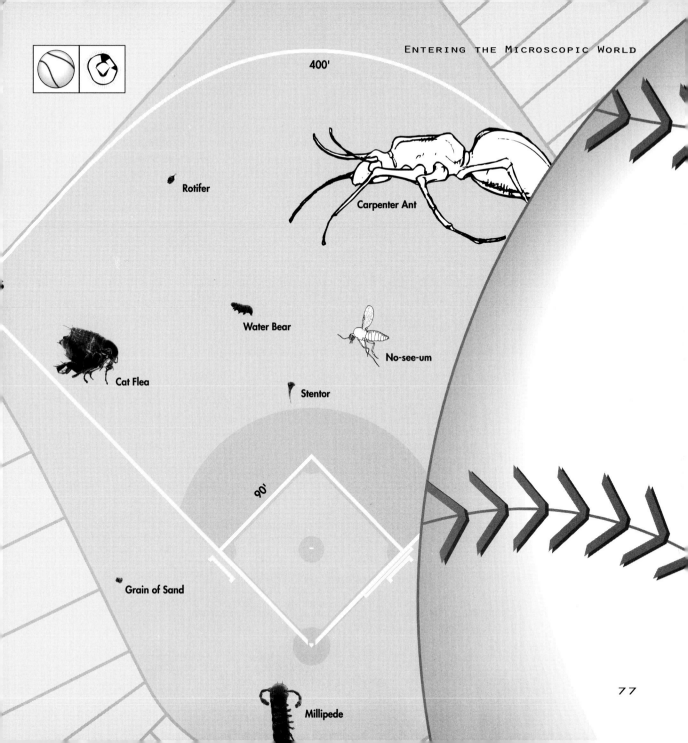

400'

Rotifer

Carpenter Ant

Water Bear

No-see-um

Cat Flea

Stentor

90'

Grain of Sand

Millipede

77

On the same scale, but zooming in on the infield, we can tour an exhibit of microscopic animals. Even single-celled organisms, like amoebas, stentors, and diatoms, are clearly visible.

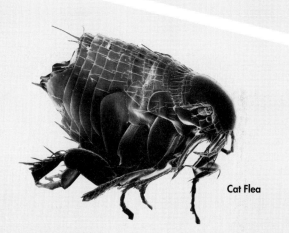

Cat Flea

Stentors are extraordinarily versatile, though they are composed of a single cell. Their flute-shaped "mouths" are wreathed with hairs that beat in rhythm, creating a miniature whirlpool that draws edible particles into their digestive cavities.

Rotifers are among the tiniest of multicelled animals. They have tiny hairs that propel them through the water. Varieties are found from Antarctic waters to tropical ponds.

A BASEBALL IS THE SIZE OF A BASEBALL PARK

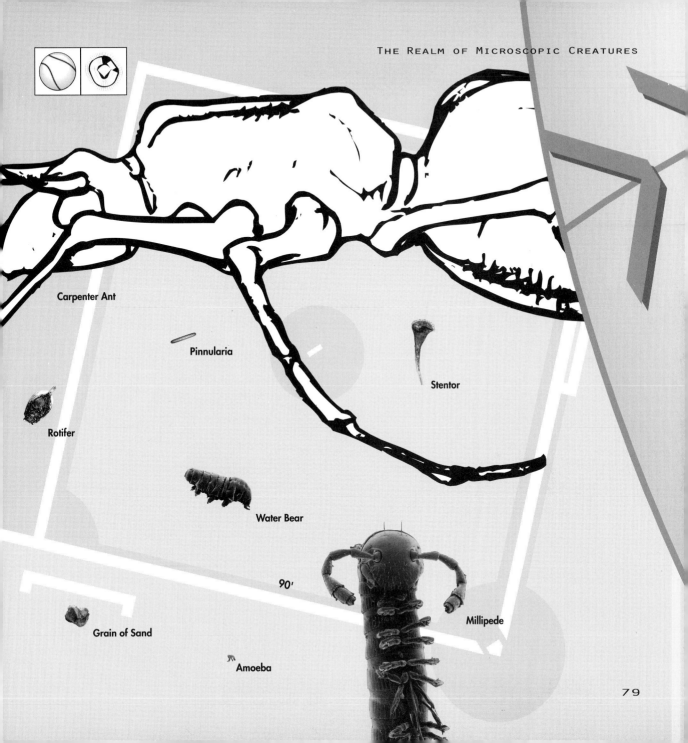

Carpenter Ant

Pinnularia

Stentor

Rotifer

Water Bear

90'

Grain of Sand

Millipede

Amoeba

79

Grain of Sand

The sperm attending the ovum were survivors of a long-distance race of half a day or more through the uterus and the fallopian tube. Several hundred million competitors dropped out along the way.

Zooming in over home plate, we enter the realm of cells. The ovum is huge compared to the tiny sperm cells seeking to penetrate it. White and red blood cells are of intermediate size, smaller than single-celled self-sufficient organisms such as the pinnularia and the peridinium.

Red and white blood cells are approximately the same size. They have entirely different functions: The red cells are the carriers of oxygen to all parts of the body. The white cells attack infectious organisms, particularly bacteria.

The pinnularia shown here is a single-celled photosynthetic alga, a member of a group known as diatoms, which is composed of as many as 10,000 species. The beautifully patterned cell walls of diatoms fit together like a box and its lid.

Arbor of Thorn Cell

A BASEBALL IS THE SIZE OF A BASEBALL PARK

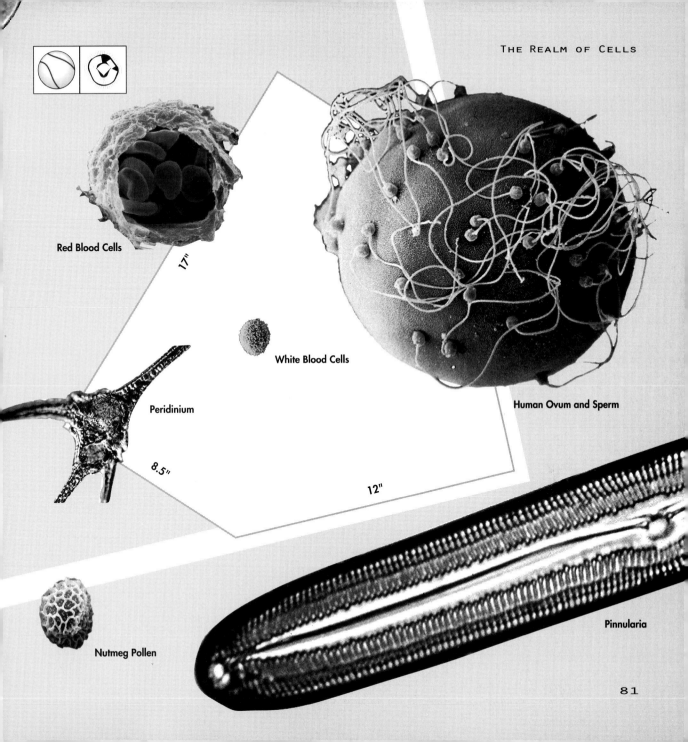

Red Blood Cells

White Blood Cells

Peridinium

Human Ovum and Sperm

17"

8.5"

12"

Nutmeg Pollen

Pinnularia

THE REALM OF MOLECULES

By expanding a baseball to the size of a baseball park, we were able to enter the microscopic world. The simplest one-celled animals and other forms of microscopic life were expanded up to several inches in length. Living cells, the "building blocks of life," became easily visible to the naked eye.

To look within the cells and examine the most basic elements of life, we must expand our baseball a great deal more — so much that it's the size of the Earth.

ASSUME A BASEBALL IS THE SIZE OF THE EARTH

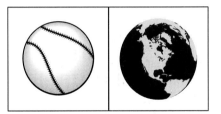

With a baseball the size of the Earth, we view the microscopic
world from orbit over the United States. A grain of sand is
smaller than Rhode Island. A cat flea spans West Virginia. Tiny
pond creatures are the size of large cities.

A BASEBALL IS THE SIZE OF THE EARTH

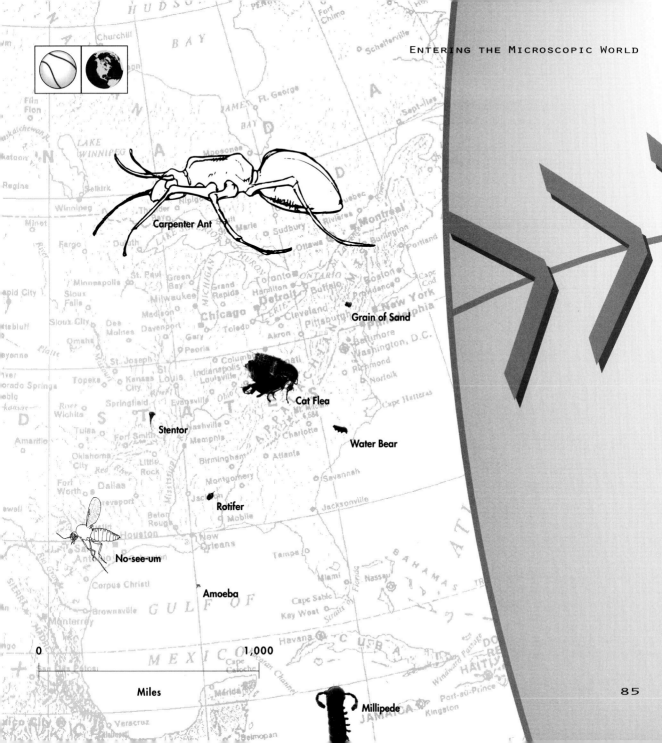

Carpenter Ant

Grain of Sand

Cat Flea

Stentor

Water Bear

Rotifer

No-see-um

Amoeba

0   1,000

Miles

Millipede

85

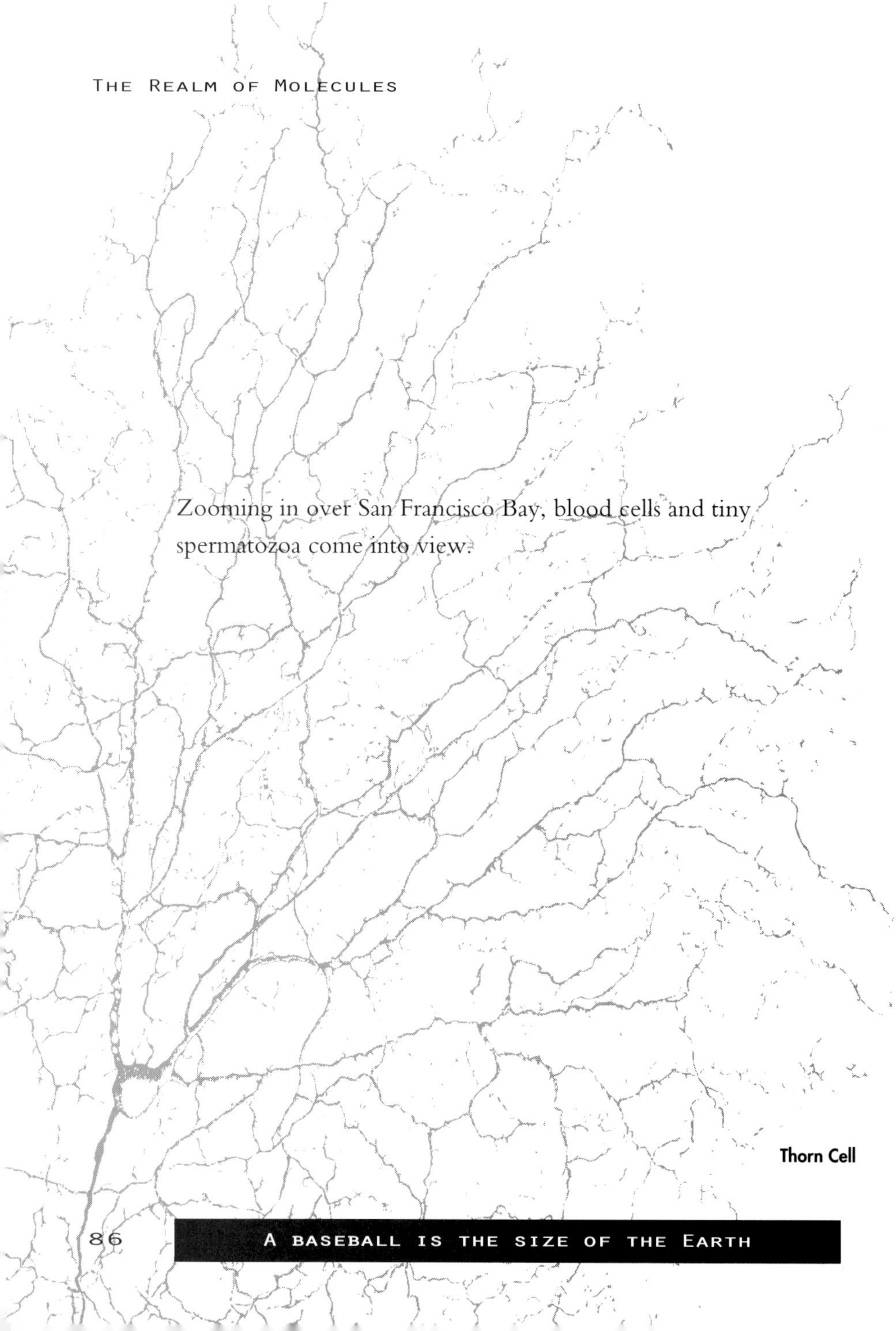

Zooming in over San Francisco Bay, blood cells and tiny spermatozoa come into view.

The thorn cell, also known as a ganglion cell, is from a primate retina. This cell receives electrical signals generated by photoreceptor cells, processes them, and projects them to a region of the brain that is concerned with visual information.

Though plants cannot travel about like animals to conjugate, they have a number of effective strategies for inseminating female receptors. Pollen grains, like the one shown here, travel on the backs of bees and with the wind.

Peridiniums are members of a large group of single-celled algae known as dinoflagellates. These organisms form a large component of marine plankton.

Thorn Cell

A BASEBALL IS THE SIZE OF THE EARTH

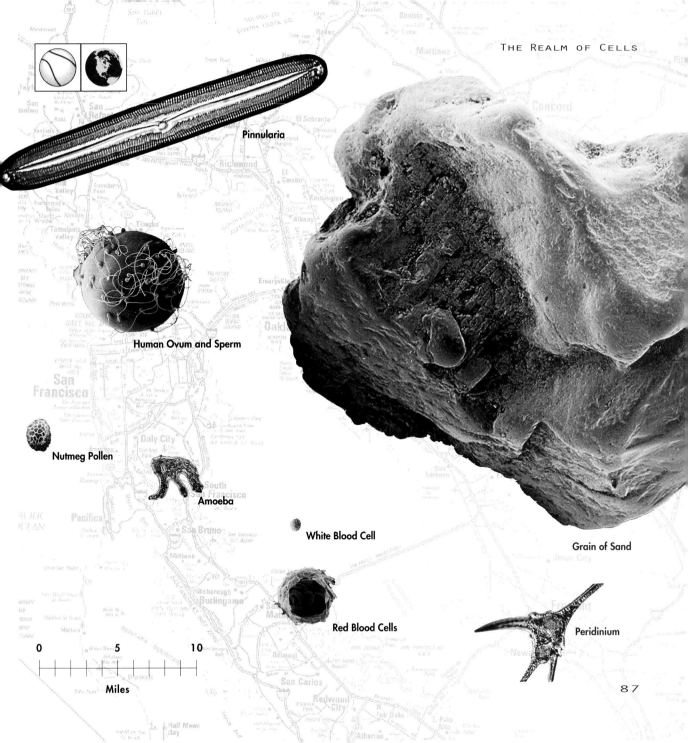

Pinnularia

Human Ovum and Sperm

Nutmeg Pollen

Amoeba

White Blood Cell

Grain of Sand

Red Blood Cells

Peridinium

0   5   10

Miles

87

Hovering over the playing field at Candlestick Park, we enter the realm of bacteria. At this scale we can see a human chromosome.

Bacteria are one-celled organisms, members of the kingdom Monera. Their cells are prokaryotic as distinguished from the eukaryotic cells of animals, plants, protists, and fungi. Prokaryotic cells have no nucleus and no organelles, the miniature organs that carry out specialized functions within the cell. A cross-section of one such organelle, a mitochondrion from a cell in a primate retina, is shown here. The organelle is much smaller than the whole cell, which would more than fill the page.

Chromosomes are structures of the cell nucleus that contain the cell's DNA and carry genetic information in the form of genes. Every human cell contains 46 chromosomes, 23 of paternal origin and 23 of maternal origin.

The enormous and dangerous-looking corkscrew spreading across the middle of the facing page is a type of bacterium called a spirochete. This one causes Lyme disease.

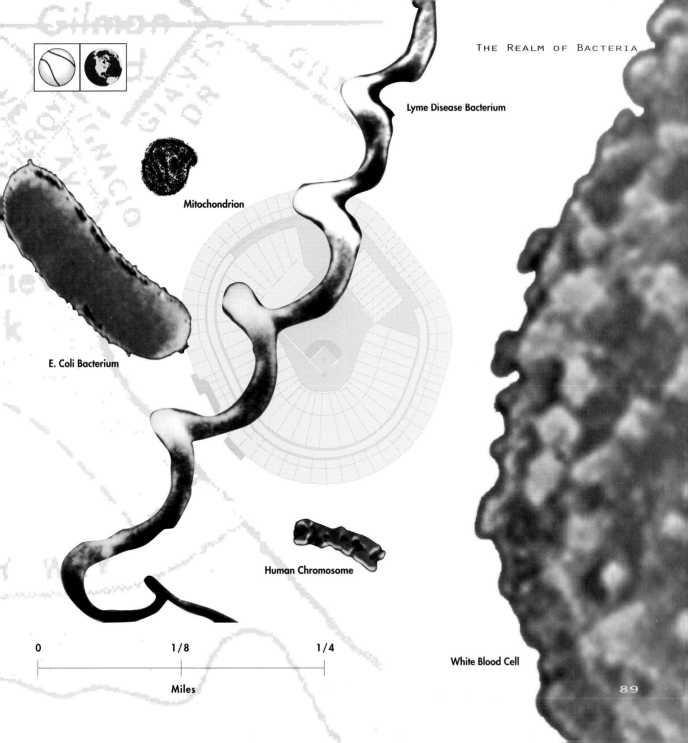

Lyme Disease Bacterium

Mitochondrion

E. Coli Bacterium

Human Chromosome

White Blood Cell

0		1/8		1/4

Miles

The status of viruses as life forms is marginal, since they are incapable of reproducing except through the agency of a living organism. In this function, as we know too well, they are highly successful.

The adenovirus is among those responsible for the common cold. The Epf1 virus may be the longest yet discovered. At this scale it would run along the bottom of five pages in a row. Its actual length is about the width of a tine of a dinner fork. The virus is not visible to the naked eye, however: it is far too thin. In fact it would not be visible under a conventional microscope, being thinner than the wavelengths of visible light.

Viruses are not much bigger than many large organic molecules such as the hemoglobin molecule shown here. This molecule is the oxygen-carrying agent in the blood. The one shown is from an earthworm. Oddly, it's much larger than human hemoglobin.

Zooming in on the infield we see viruses in detail, dwarfed by an E. coli bacterium. At this distance we can discern large, complex molecules.

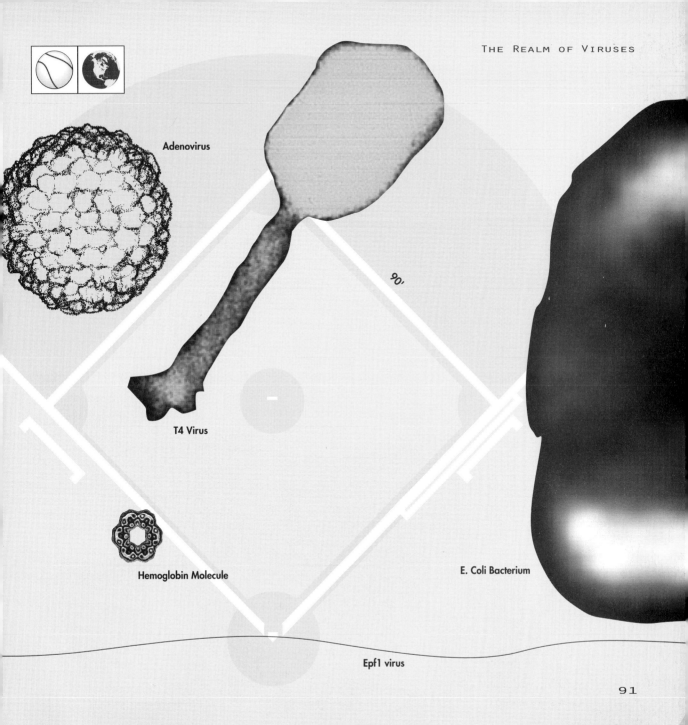

Adenovirus

90'

T4 Virus

Hemoglobin Molecule

E. Coli Bacterium

Epf1 virus

Shown above and to the right of home plate is the face of a sodium chloride crystal (table salt). The sodium and chlorine atoms bond in a way that forms the lattice structure characteristic of crystals. In the bond, electrons from sodium atoms become associated with chlorine atoms. The altered atoms are called ions. This micrograph, made with an atomic force microscope, shows only the chlorine ions. The much smaller sodium atoms are hidden in the crevices.

From a batter's view of home plate the realm of molecules is fully exposed.

The molecules above and to the left of home plate are segments of polyethylene molecules.

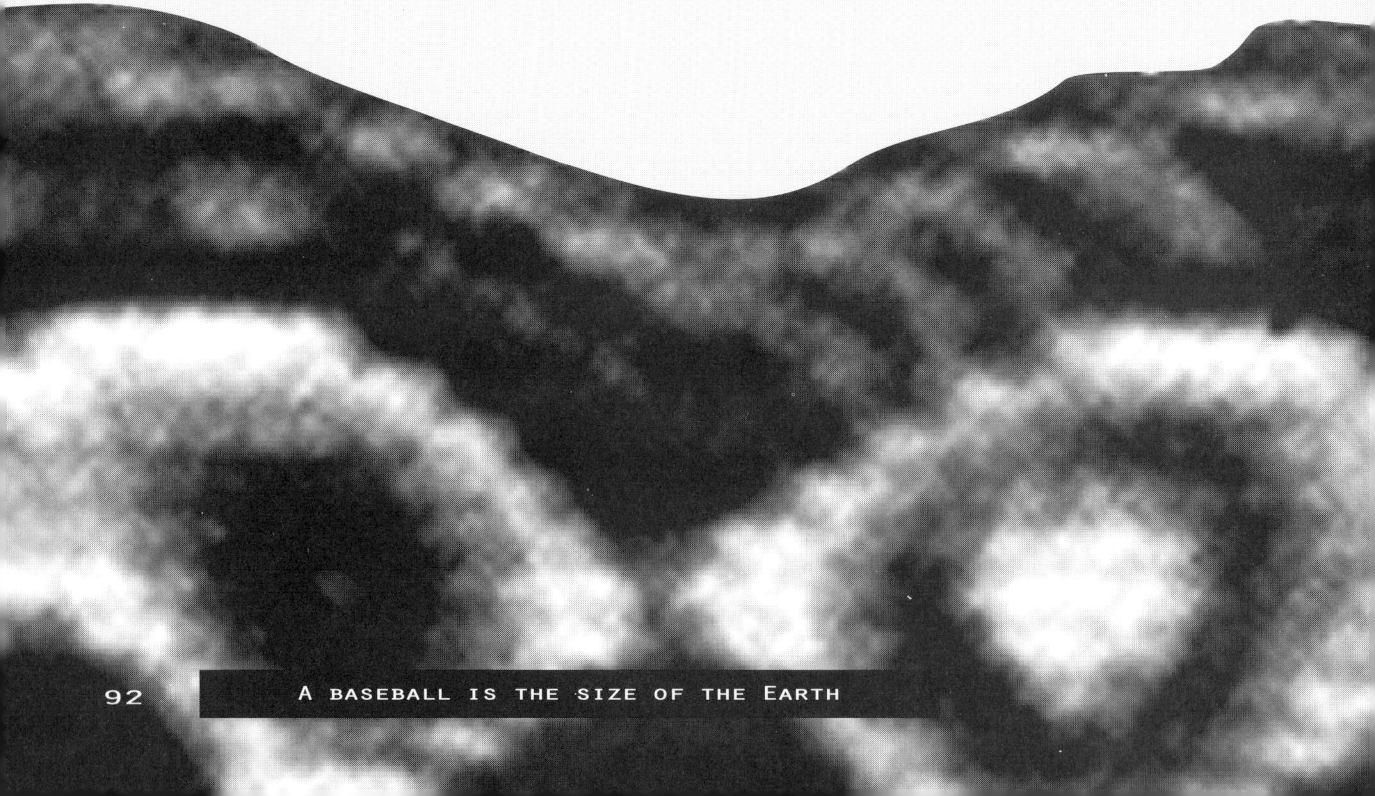

A BASEBALL IS THE SIZE OF THE EARTH

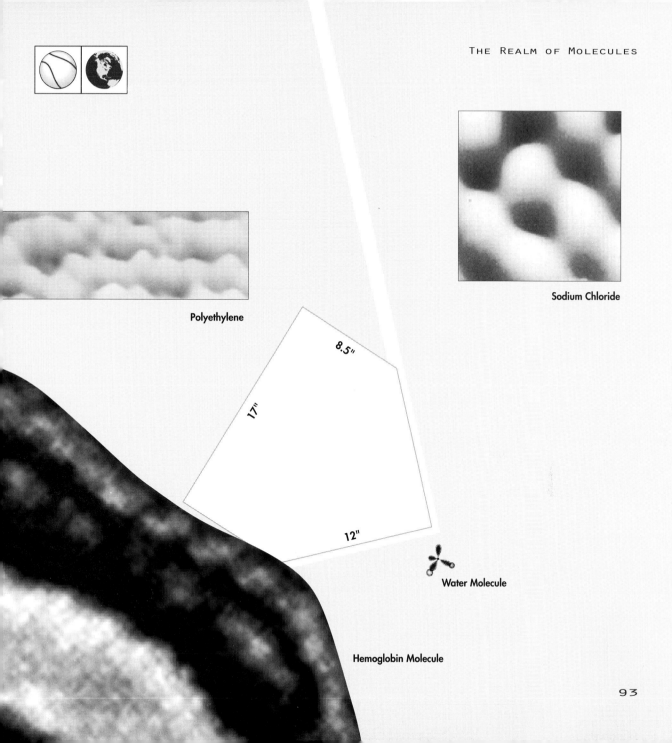

Polyethylene

Sodium Chloride

8.5"

17"

12"

Water Molecule

Hemoglobin Molecule

The Realm of Atoms

We expanded a baseball to the size of the Earth in order to render cells, bacteria, viruses, and molecules visible.

To visualize the fundamental unit of matter — the atom — we will expand a baseball a great deal more, so much that it reaches from the Earth to the moon.

ASSUME A BASEBALL REACHES FROM THE EARTH TO THE MOON

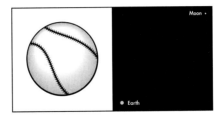

Moon ·

· Earth

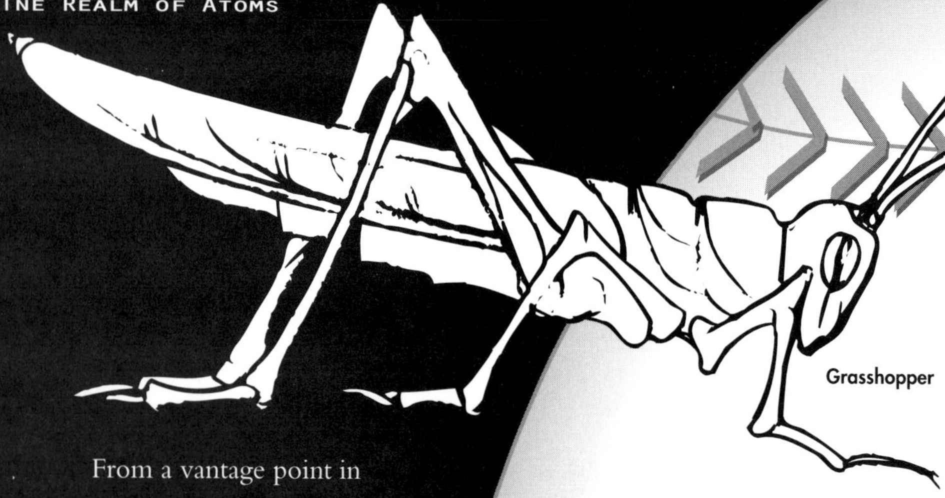

Grasshopper

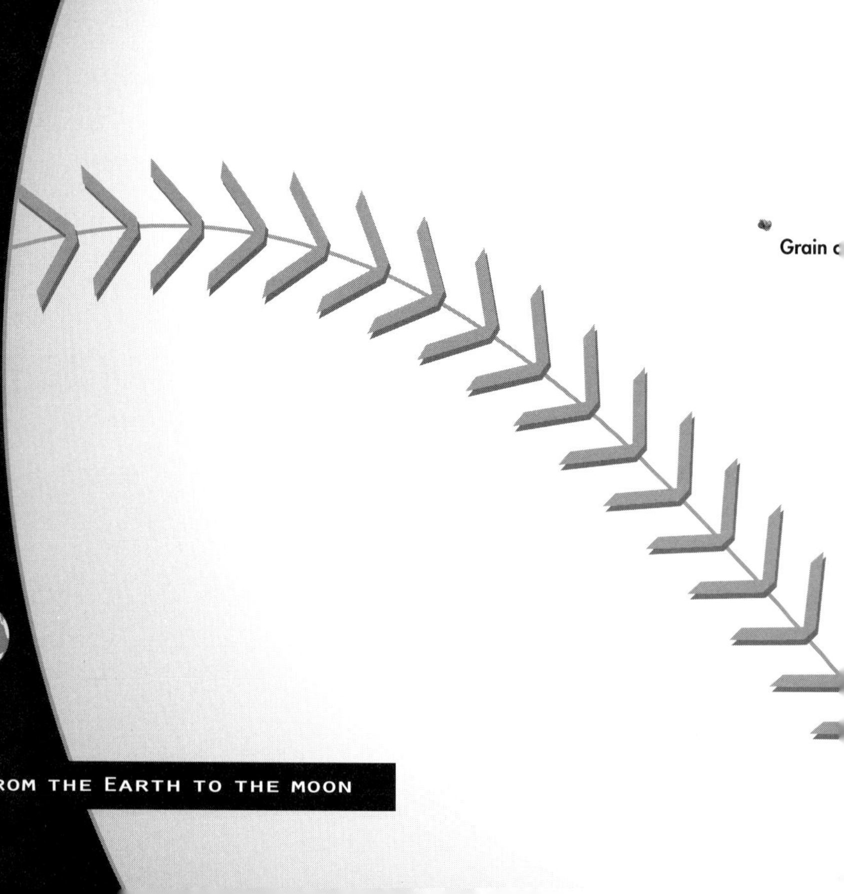

From a vantage point in
outer space we see a baseball
expanded so that it spans the
distance from the Earth and
the moon. On this scale a cat
flea is larger than the Pacific
Ocean, a rotifer is as large
as Brazil.

Grain c

Earth

A BASEBALL REACHES FROM THE EARTH TO THE MOON

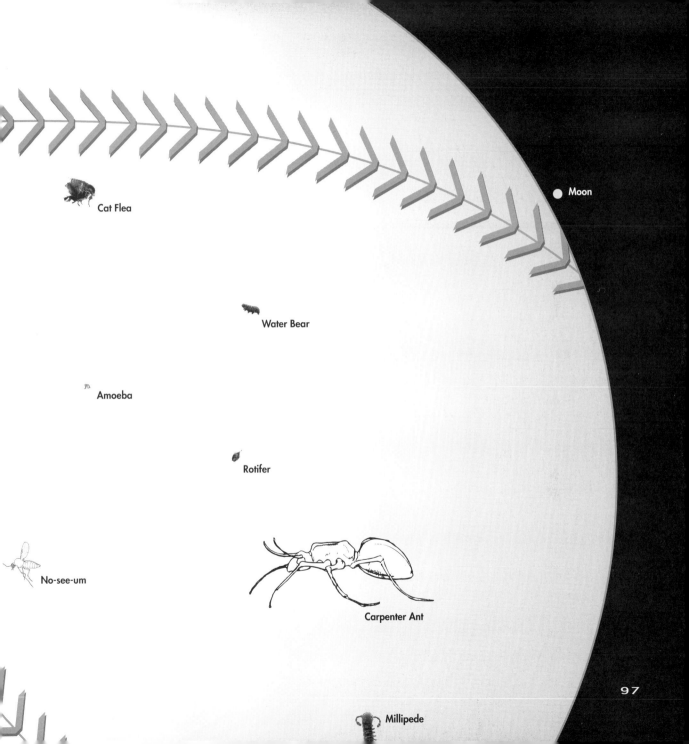

Cat Flea

Water Bear

Amoeba

Rotifer

No-see-um

Carpenter Ant

Moon

Millipede

97

The range of sizes of creatures is controlled by the laws of physics. Organisms without skeletons, such as rotifers and stentors, cannot support much weight. This is not much of a problem for water dwellers — the water holds them up. Even so, they could not grow very large or they would be torn apart by waves.

Insects have external skeletons, a system that works very well on a small scale but would be highly inefficient at larger scales, limiting their body size to about that of a grass frog.

The limits for vertebrates are much more tolerant, to which the dinosaurs attest. But the requirements of supporting a huge frame against gravity impose problems, which increase with size. Thus it is a marine creature, the blue whale, that evolved to become the largest animal that ever lived.

Moving close to Earth, we observe a grain of sand larger than Alaska. Single-celled life forms come into view.

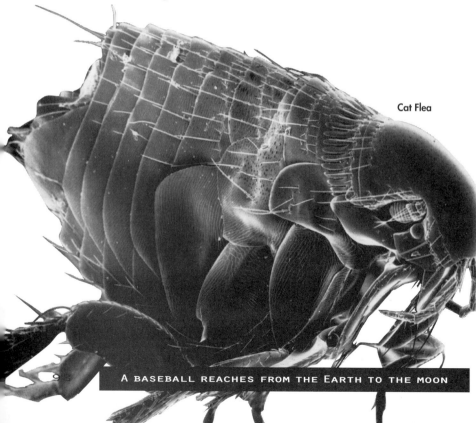

Cat Flea

A BASEBALL REACHES FROM THE EARTH TO THE MOON

Moon
• Earth

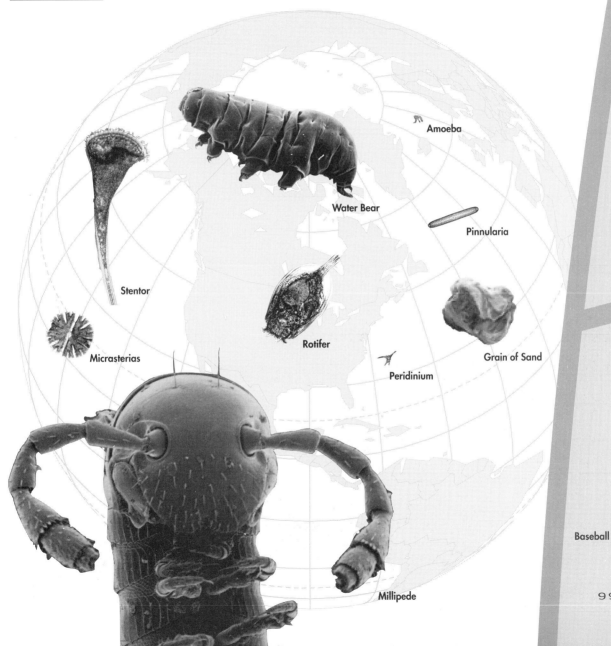

Amoeba

Water Bear

Pinnularia

Stentor

Rotifer

Grain of Sand

Micrasterias

Peridinium

Baseball

Millipede

99

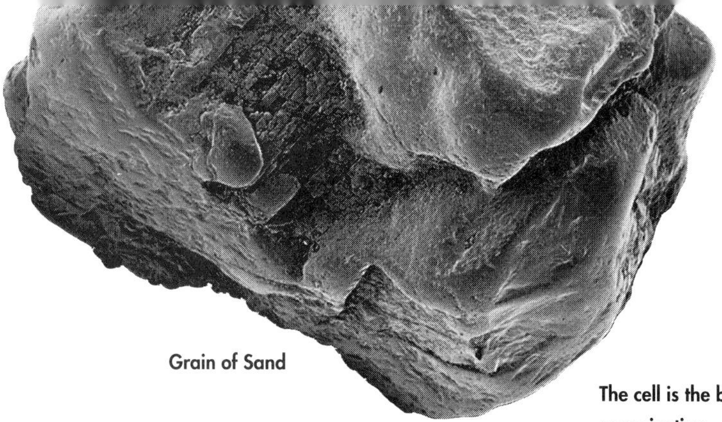

Grain of Sand

The cell is the basic unit of organization — the smallest form of organization of self-perpetuating life.

The single cell of organisms such as amoebas, pinnularias, peridiniums, and the microsterias shown here is sufficiently versatile to perform all the functions necessary to sustain and propagate life. By contrast, the cells within our bodies and those of other multicelled organisms perform highly specialized functions.

The microsterias is a desmid, a single-celled freshwater green alga. Unlike the peridinium, it is without means of propelling itself through the water.

Zooming in over the western United States we can see the smallest cells.

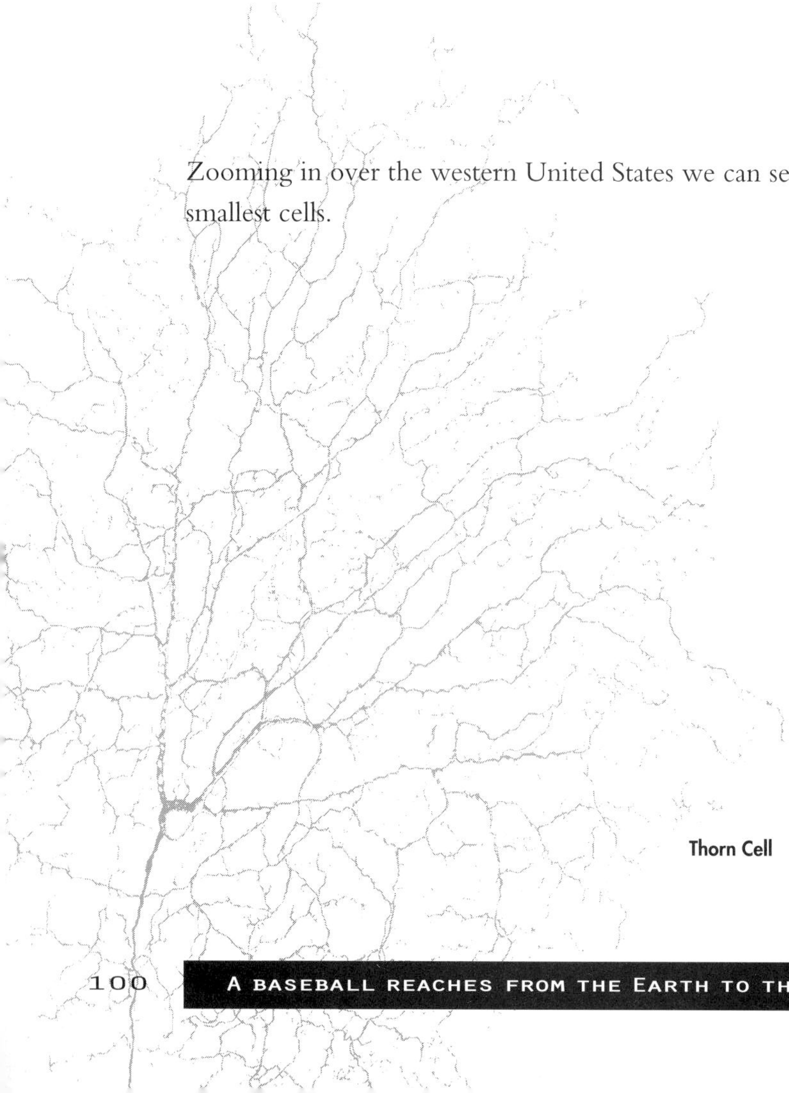

Thorn Cell

A BASEBALL REACHES FROM THE EARTH TO THE MOON

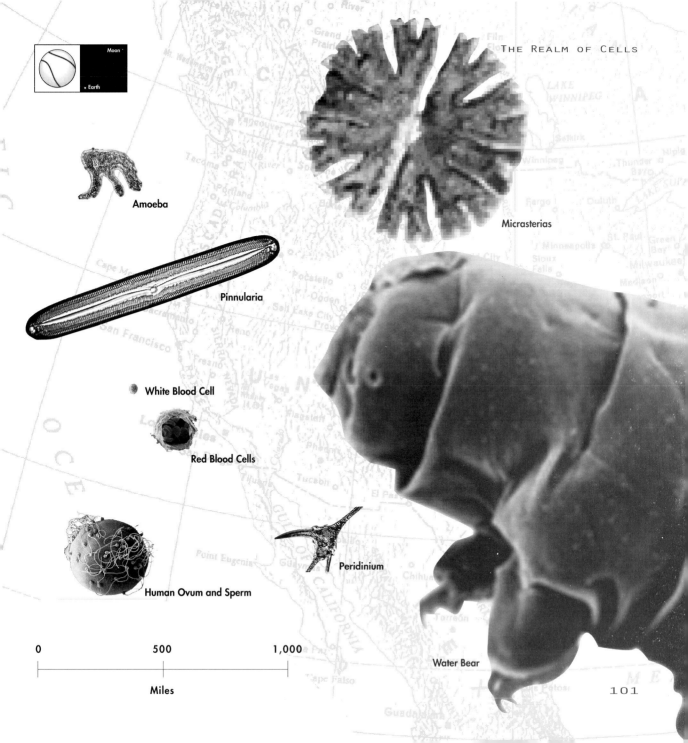

Moon
• Earth

Amoeba

Micrasterias

Pinnularia

White Blood Cell

Red Blood Cells

Peridinium

Human Ovum and Sperm

Water Bear

0   500   1,000

Miles

Represented here are the three
common shapes of bacteria —
rods, spirals, and spheres.

This electron micrograph image
of an E. coli bacterium shows its
DNA extruded. The bacterium
appears to be no more than a
container for its DNA. Actually it
contains some 5,000 different
kinds of molecules besides DNA.

The DNA strand is about
600,000 times longer than it is
wide. If drawn out in a straight
line it would be 700 times the
length of the entire cell.

Keeping in mind that a baseball reaches from the Earth to the
moon and that a grain of sand is larger than Alaska, we zoom in
over the Bay area to view the realm of bacteria.

A BASEBALL REACHES FROM THE EARTH TO THE MOON

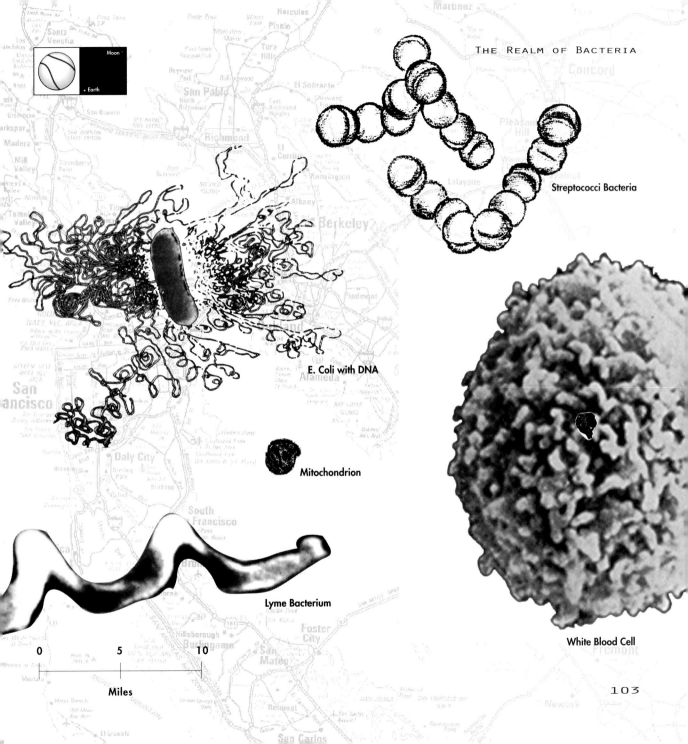

Streptococci Bacteria

E. Coli with DNA

Mitochondrion

Lyme Bacterium

White Blood Cell

Moon

Earth

0   5   10

Miles

103

In nature there is generally a correlation between power and size. Little fish do not normally attack big ones. Jackals do not attack elephants, nor, for that matter, do lions. In the microscopic world size is of less importance. Chemical weapons and defenses are what count. So, certain bacteria are forever on the move among the cells of larger organisms, searching for an opportunity to invade. And bacteria themselves are subject to attack by much smaller viruses.

If successful, the T4 virus will insert its spike into the E. coli. It not only gains nourishment from its victim, it also seeks to use the larger organism's enzymes and metabolic processes to reproduce itself. Like the invading pod people in the movie, the T4 is a "body snatcher."

Closer still, we see bacteria that span the city and viruses the size of Candlestick Park.

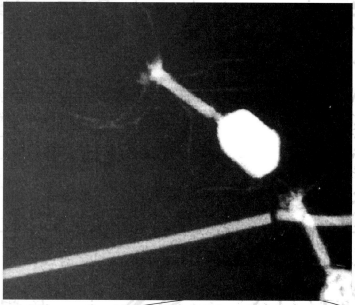

T4 Virus
Tobacco Mosaic Virus (Rod Shaped)

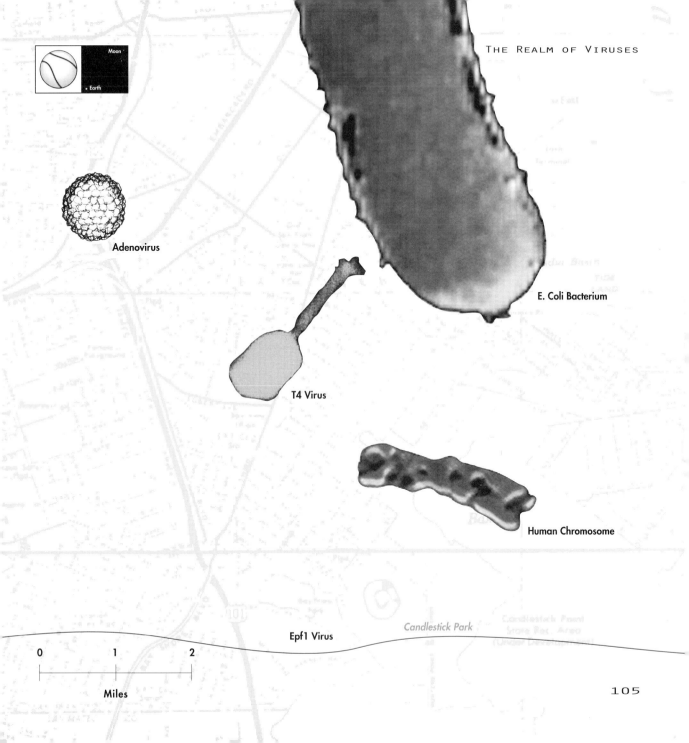

Moon

Earth

Adenovirus

E. Coli Bacterium

T4 Virus

Human Chromosome

Epf1 Virus

Candlestick Park

0   1   2

Miles

Polymers are large molecules, often long chains of atoms bonded together. Some chains align to form crystals, as shown in the image on the far right. As can be seen, one chain has folded to make a hairpin turn.

The water molecule is composed of two hydrogen atoms and an oxygen atom. This combination has formed because the oxygen atoms "prefer" ten electrons but only have eight, and hydrogen atoms "prefer" two electrons but only have one. The nature of these "preferences" is highly technical. Suffice to say here that the atoms satisfy their yearnings by sharing electrons.

The core of the water molecule is the nucleus of the oxygen atom, which is sixteen times more massive than the nucleus of the hydrogen atom.

DNA Strand

Hovering over Candlestick Park — on a scale where a blood cell would be as big as the San Francisco Bay area and a bacterium would span the city, we find molecules, some the size of the playing field, others about the size of home plate. The nucleus of a cell might be about the size of the ball park.

Moon

· Earth

90'

Portion of Polyethylene Chain

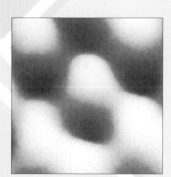

Sodium Chloride Crystal Lattice

Water Molecule

Hemoglobin Molecule

107

As we approach home plate, small molecules and large atoms are revealed.

Atoms and their characteristics can be quite precisely described in words and mathematical notations, but producing a "picture" of one is somewhat less of a science and more of an art. The electron micrograph of a single atom of gold reproduced here might be mistaken for a smudge, which in a way is a legitimate depiction, because the particles of which the atom is composed are vibrating at tremendous speed.

In principle, an atom cannot be seen — it is much smaller than the wavelength of visible light. Similarly, the electrons in the water molecule can only be depicted as a cloud whose density and shape are correlated to probabilities of their locations.

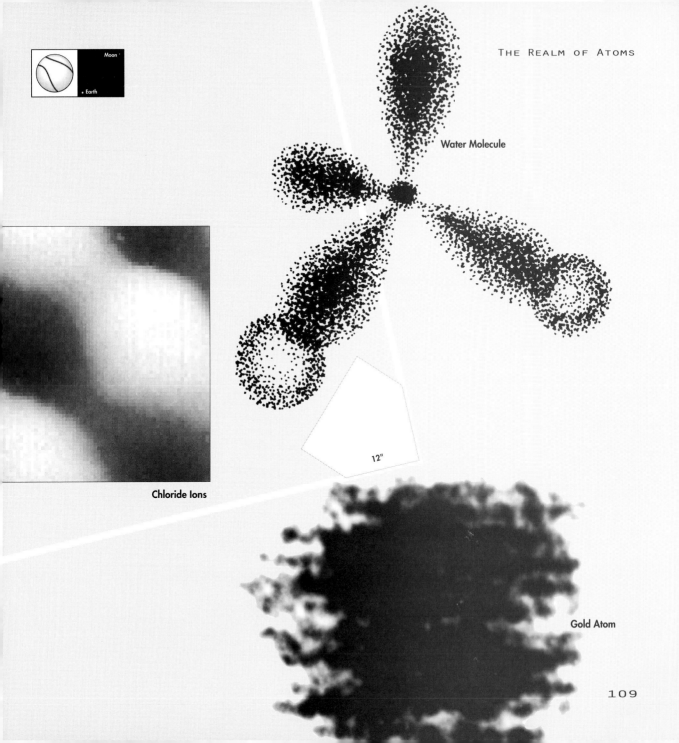

Moon

Earth

Water Molecule

12"

Chloride Ions

Gold Atom

The Realm of Elementary Particles

We have expanded a baseball so much that it would reach from the Earth to the moon, and zoomed in past one-celled organisms as big as small states, blood cells larger than cities, viruses the size of ball parks, molecules as big as an infield, and atoms larger than home plate.

Although the atom is the basic unit of matter, the atom itself is virtually empty. Its "size" is only a measure of the space in which its constituent electrons may be found. The atom's nucleus is far smaller than the atom itself, so small that to visualize it we must expand a baseball even more — so much that it would reach from the Earth to the sun.

ASSUME A BASEBALL REACHES FROM THE EARTH TO THE SUN

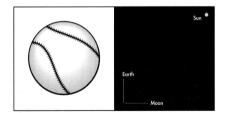

Earlier in this book we examined how far the Earth is from the sun. If the sun were the size of the tiny disc in the upper right-hand corner, the Earth and the moon would each be microscopic. The arrowheads point to their locations. The distance between the points of the arrows represents the distance between the Earth and the moon. The distance shown to the sun is on the same scale.

Earth

Moon →

• Sun

The hydrogen atom is the smallest and simplest atom. A drawing of a cross-section of it is shown here.

The proton that forms the nucleus of the atom is too small to be seen. The electron, being far smaller than the proton, if it can be said to have any size at all, cannot be seen either. Nor would we be able to locate it. The myriad of dots surrounding the nucleus represents only probabilities of where the electron might be at a given instant. The swath of the orbit within which the electron is likely to be 90 percent of the time may be said to define the atom's size.

Gold is one of the denser elements, as the weight of a gold coin attests. Whereas the nucleus of a hydrogen atom has only a lone proton and usually no neutrons, the nuclei of gold atoms have 79 protons and, on average, 118 neutrons. Nonetheless an atom of gold, like all atoms, is mostly empty space. Even enlarged to this scale, none of the particles comprising it are visible. We see only the cloud produced by their active presence.

By expanding a baseball so much that it spans the distance from Earth to the sun, we are able to fit a hydrogen atom inside Candlestick Park. From this vantage point, the atom's nucleus, located over home plate, is too small to see.

Gold Atom

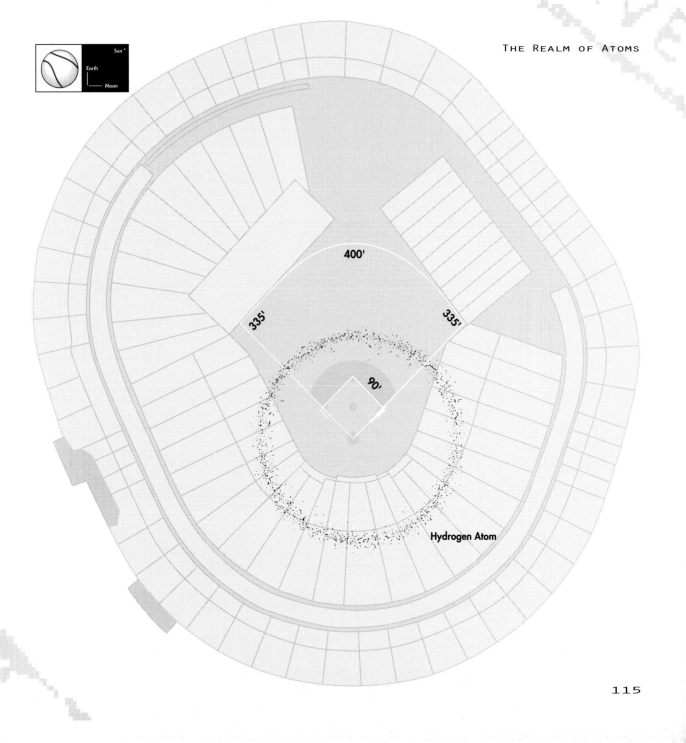

Sun *
Earth
Moon

400'

335'

335'

90'

Hydrogen Atom

Only by zooming in over home plate can we see the nucleus of the hydrogen atom: its single proton.

The proton — the core of the hydrogen atom — exists in solitude. The atom's electron is forced to remain at a "great distance" confined more or less to its orbital path. The proton, like other elementary particles, is in constant motion. Its position at any moment cannot be precisely determined, but scientists know it is there. When a particle accelerator propels an electron in its vicinity, the electron is deflected, revealing something about this elemental piece of matter that no one will ever see.

Much smaller than the proton are its mysterious components, called quarks. Their size is unmeasured and unmeasurable. They are known only by their effects.

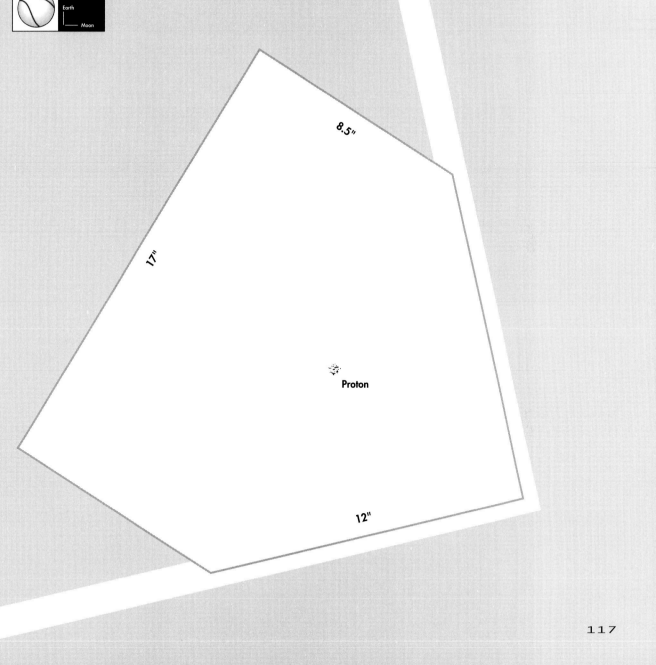

PART III

LOOKING BACKWARD AND LOOKING FORWARD

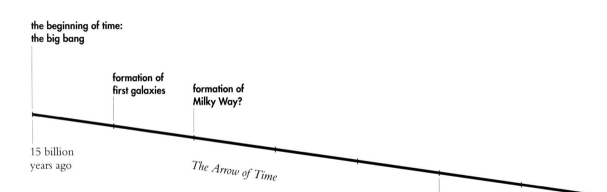

the beginning of time:
the big bang

formation of
first galaxies

formation of
Milky Way?

15 billion
years ago

The Arrow of Time

10 billion
years ago

How can we grasp the vast amount of time that has elapsed since the beginning of the universe? How can we compare it to the years that comprise a human life span? Begin by thinking of time as a moving object.

Imagine time moves one mile every one million years — 1,000 miles every billion years. Where did time begin in relation to the moment you are reading this book? If the universe is 15 billion years old, time began 15,000 miles out in space.

This illustration shows the arrow of time moving from its beginning 15 billion years ago until 4.5 billion years ago, when the Earth was formed.

Galaxies and Stars

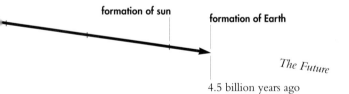

formation of sun

formation of Earth

The Future

4.5 billion years ago

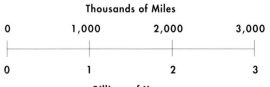

Thousands of Miles

| 0 | 1,000 | 2,000 | 3,000 |

| 0 | 1 | 2 | 3 |

Billions of Years

Time Past: The Beginning of the Universe to the Formation of Earth

Time continues from the Earth's formation, 4.5 billion years ago. We follow it as it moves onward until a billion years ago.

Life begins relatively soon after the formation of the Earth. During the next three billion years (3,000 miles on our scale) it develops into many different forms, none of them large enough to be seen with the naked eye.

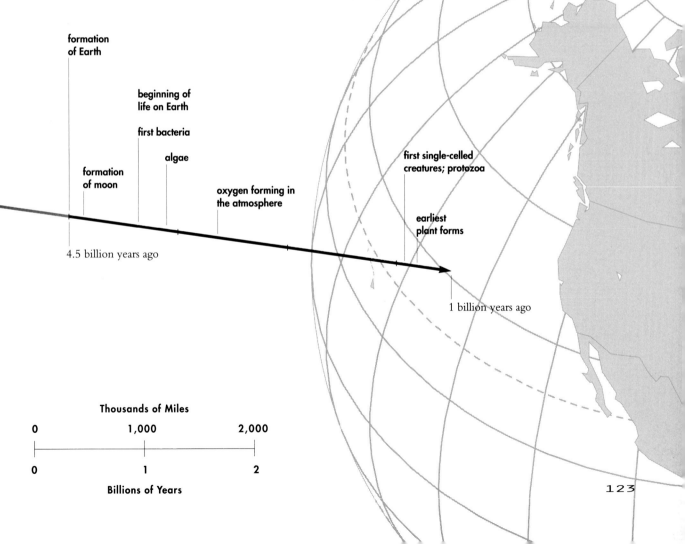

ONE MILE = ONE MILLION YEARS

Great Variations in Climate and Topography

formation
of Earth

beginning of
life on Earth

first bacteria

algae

formation
of moon

first single-celled
creatures; protozoa

oxygen forming in
the atmosphere

earliest
plant forms

4.5 billion years ago

1 billion years ago

Thousands of Miles

0	1,000	2,000

0	1	2

Billions of Years

123

LOOKING BACKWARD

Time Past: The Appearance of Life

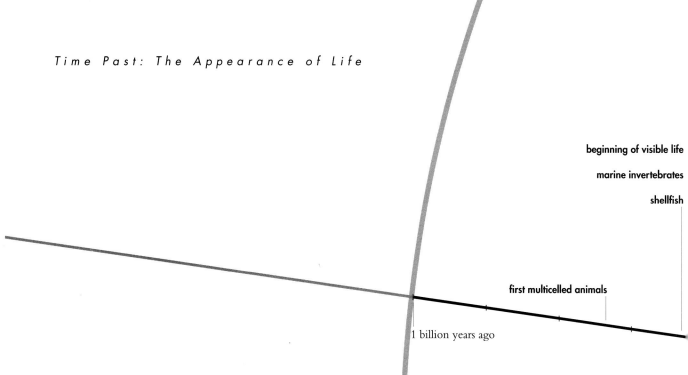

beginning of visible life

marine invertebrates

shellfish

first multicelled animals

1 billion years ago

Time travels across the Pacific Ocean. The first multicelled animals appear. Before the arrow of time reaches within five hundred miles of the California coast, a startling variety of life forms appears. Fish begin populating the oceans. Soon afterward the first protists and plants arrive on land, followed by animals: amphibians, reptiles, dinosaurs, primitive mammals, and birds. Earlier established life forms continue to evolve even as new ones appear.

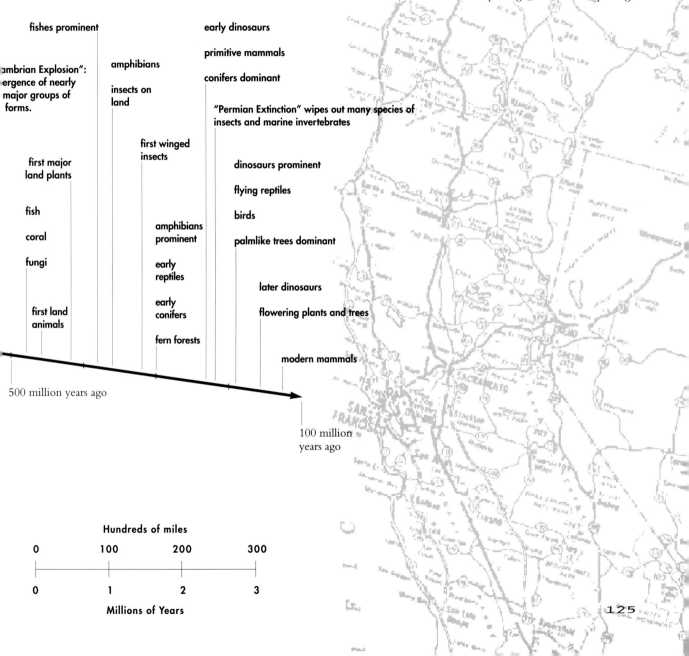

ONE MILE = ONE MILLION YEARS

fishes prominent

early dinosaurs

primitive mammals

ambrian Explosion":
ergence of nearly
major groups of
forms.

amphibians

conifers dominant

insects on
land

"Permian Extinction" wipes out many species of
insects and marine invertebrates

first winged
insects

first major
land plants

dinosaurs prominent

flying reptiles

fish

birds

coral

amphibians
prominent

palmlike trees dominant

fungi

early
reptiles

early
conifers

later dinosaurs

first land
animals

flowering plants and trees

fern forests

modern mammals

500 million years ago

100 million
years ago

Hundreds of miles

0	100	200	300

0	1	2	3

Millions of Years

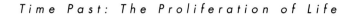

Time Past: The Proliferation of Life   *Continuing Fluctuati*

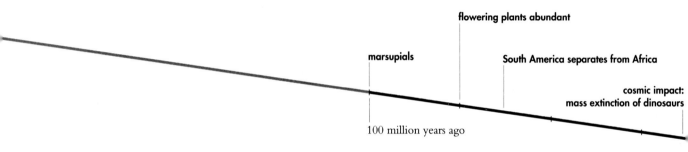

flowering plants abundant

marsupials

South America separates from Africa

cosmic impact:
mass extinction of dinosaurs

100 million years ago

We follow the arrow of time for ninety million years as it
approaches the California coast, a period marked by the
extinction of the dinosaurs, the proliferation of mammals, and
the appearance of the first primates. The later half of this period
brings the arrival of most modern animals and the extinction of a
great many others.

Climate and Topography

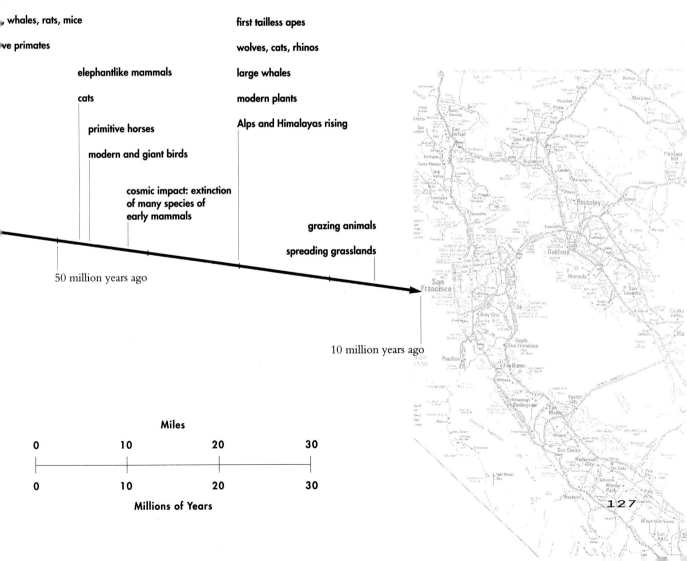

ammalian carnivores

whales, rats, mice

ve primates

elephantlike mammals

cats

primitive horses

modern and giant birds

cosmic impact: extinction
of many species of
early mammals

first tailless apes

wolves, cats, rhinos

large whales

modern plants

Alps and Himalayas rising

grazing animals

spreading grasslands

50 million years ago

10 million years ago

Miles

| 0 | 10 | 20 | 30 |

| 0 | 10 | 20 | 30 |

Millions of Years

127

Time Past: The Age of Mammals

Mountain Building; Glac

common human-chimpanzee ancestor

10 million years ago

The arrow of time reaches the coast and travels through the city of San Francisco.

The first hominids, species looking eerily similar to our own and with progressively bigger brains than their ape ancestors, appear in Africa.

THE APPEARANCE OF HOMINIDS

10 million years ago to 1 million years ago

vances and Retreats .
fes
os
baboons
first hominids?
modern camels
bears
large carnivores
Australopithecus aferensis: bipedal prehumans
Australopithecus africanus
Homo habilis: increase in brain size; stone tools
Homo erectus: further increase in brain size; specialized tools
first use of fire?
5 million years ago
1 million years ago
Candlestick Park
Miles
0
1
2
3
0
1
2
3
Millions of Years

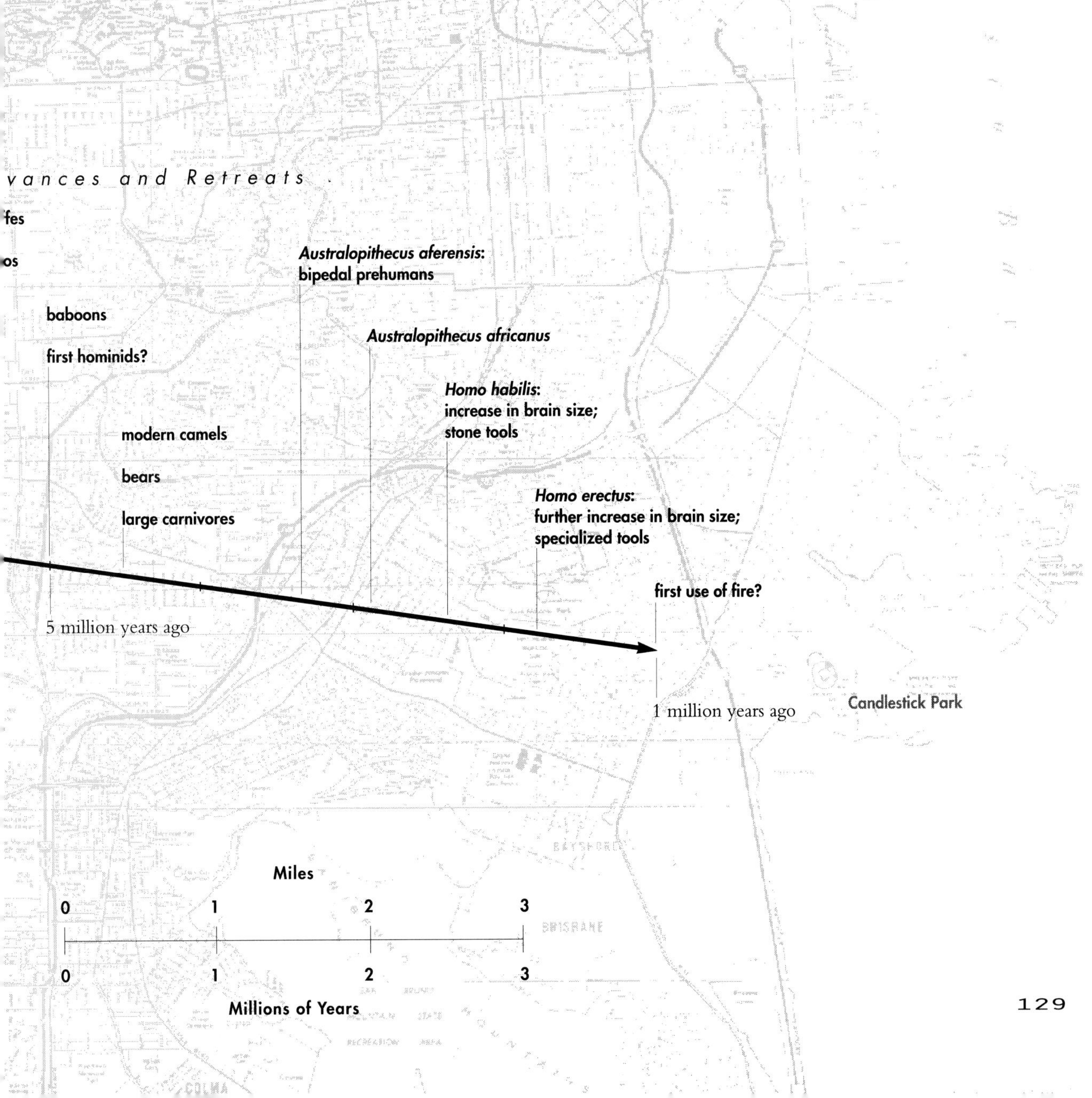

Time Past: The Appearance of Hominids

proliferation of *Homo erectus*

1 million years ago

The arrow of time reaches within a few blocks of Candlestick Park. Hominids resembling modern man appear. With the entry of time into the stadium come the first anatomically modern humans.

1 million years ago to 100,000 years ago

Recurring Ice Ages: Extinction of Many Large Animals and Birds

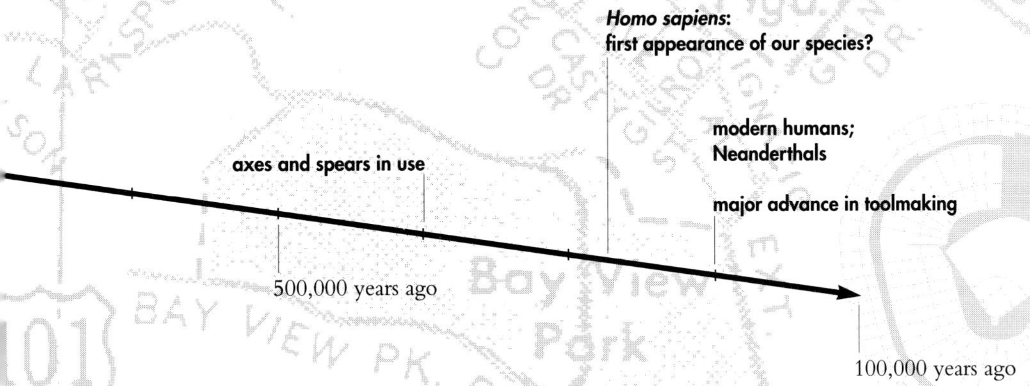
Homo sapiens:
first appearance of our species?
modern humans;
Neanderthals
axes and spears in use
major advance in toolmaking
500,000 years ago
100,000 years ago

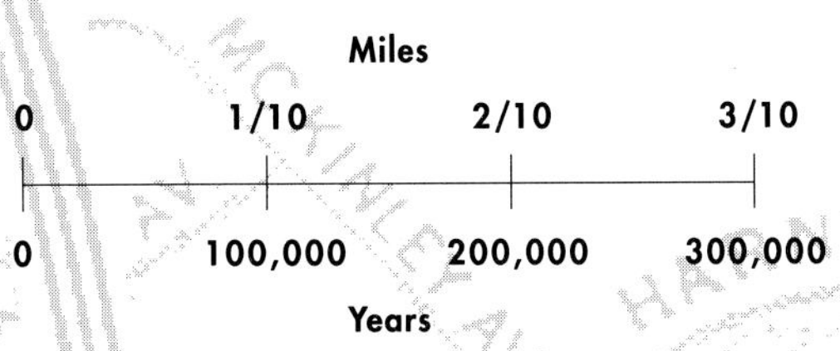
Miles
0
1/10
2/10
3/10
0
100,000
200,000
300,000
Years

Time Past: The Appearance of Humans

Extinction of Ma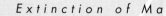

early knives

Neanderthals dominate Euro

stone lamps fueled by anima

modern humans ascendant

100,000 years ago

Time proceeds across the stadium. Humans begin spreading throughout the world. The human advantage over other large species becomes decisive. As time reaches the infield, humans are cultivating crops and making clothes and highly specialized implements. The stage is set for the beginning of civilization.

ONE MILE = ONE MILLION YEARS

...ecies at the Hands of Humans

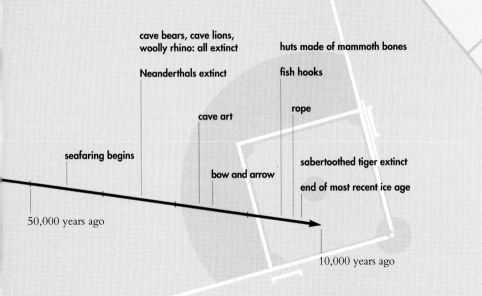

cave bears, cave lions,
woolly rhino: all extinct

huts made of mammoth bones

Neanderthals extinct

fish hooks

cave art

rope

seafaring begins

sabertoothed tiger extinct

bow and arrow

end of most recent ice age

50,000 years ago

10,000 years ago

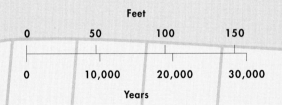

Feet

0	50	100	150

0	10,000	20,000	30,000

Years

Time Past: The Ascendancy of Humans

Late Stone

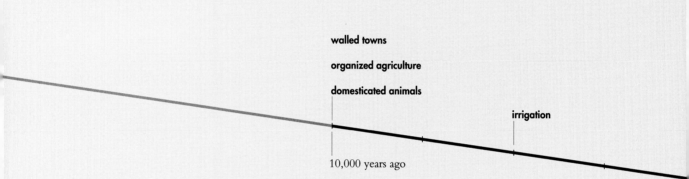

walled towns

organized agriculture

domesticated animals

irrigation

10,000 years ago

After traveling almost 30,000 miles through space, across thousands of miles of ocean, across the city of San Francisco and most of the ball field at Candlestick Park, the arrow of time approaches home plate. The events recorded in history books unfold.

THE RISE OF CIVILIZATIONS

10,000 years ago to 1,000 years ago

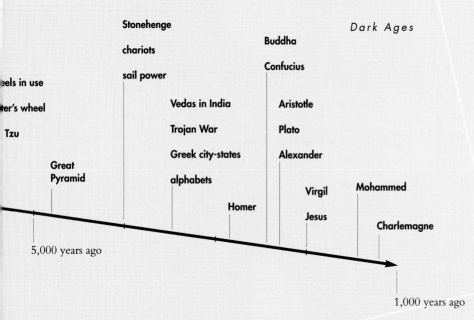

Bronze Age

Iron Age

Stonehenge

chariots

sail power

Buddha

Confucius

Dark Ages

Vedas in India

Trojan War

Greek city-states

alphabets

Aristotle

Plato

Alexander

eels in use

ter's wheel

Tzu

Great
Pyramid

Homer

Virgil

Jesus

Mohammed

Charlemagne

5,000 years ago

1,000 years ago

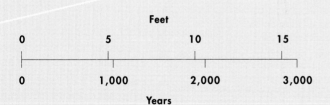

Feet

0	5	10	15

0	1,000	2,000	3,000

Years

135

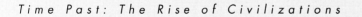

Time Past: The Rise of Civilizations

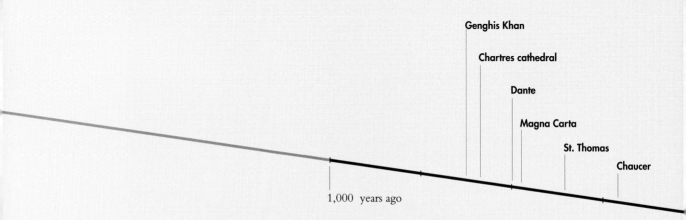

Genghis Khan

Chartres cathedral

Dante

Magna Carta

St. Thomas

Chaucer

1,000 years ago

The arrow of time moves from a thousand to a hundred years ago — a distance of just four and a half feet on our scale, yet a period of transformation of human existence and of the beginning of human transformation of the Earth.

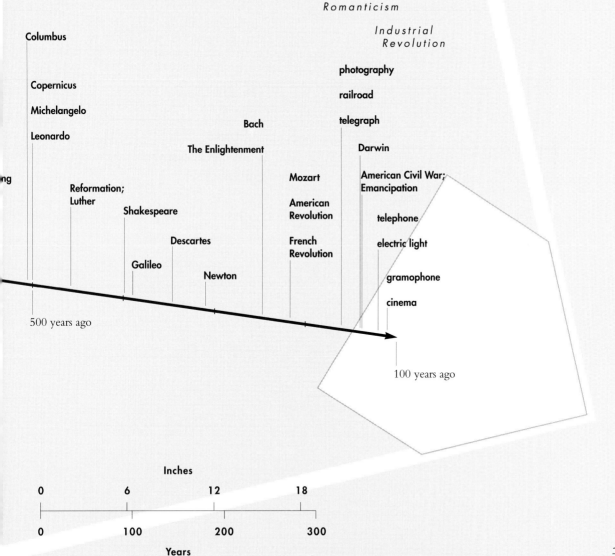

ONE MILE = ONE MILLION YEARS

Renaissance

Exploration; Colonial Conquests

Romanticism

*Industrial
Revolution*

Columbus

Copernicus

Michelangelo

Leonardo

photography

railroad

telegraph

Bach

The Enlightenment

Darwin

ng

American Civil War;
Emancipation

Reformation;
Luther

Mozart

Shakespeare

American
Revolution

telephone

Descartes

French
Revolution

electric light

Galileo

gramophone

Newton

cinema

500 years ago

100 years ago

Inches

0	6	12	18

0	100	200	300

Years

Time Past: The Rise of Technology

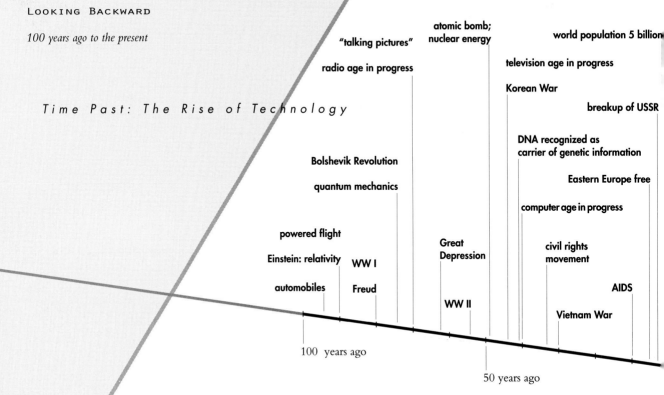

atomic bomb;
nuclear energy

world population 5 billion

"talking pictures"

radio age in progress

television age in progress

Korean War

breakup of USSR

DNA recognized as
carrier of genetic information

Bolshevik Revolution

quantum mechanics

Eastern Europe free

computer age in progress

powered flight

Great
Depression

civil rights
movement

Einstein: relativity   WW I

AIDS

automobiles   Freud

WW II

Vietnam War

100 years ago

50 years ago

th
prese

The arrow of time crosses home plate. It creeps over it even as
you look at this illustration, moving at the rate of a sixteenth of
an inch a year. The staggering events of the past hundred years
foster humility about predicting the future, but we have some
idea of the threats and opportunities that lie ahead.

ONE MILE = ONE MILLION YEARS

War Lords?

Ethnic Wars?

Famines?

Plagues?

Ecological Disasters?

Cheap, Clean Energy?

Conquest of Disease?

New Enlightenment?

World-Wide Democracy?

World Peace?

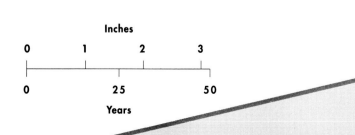

world population 10 billion?

50 years from now

100 years from now

Inches

0	1	2	3

0	25	50

Years

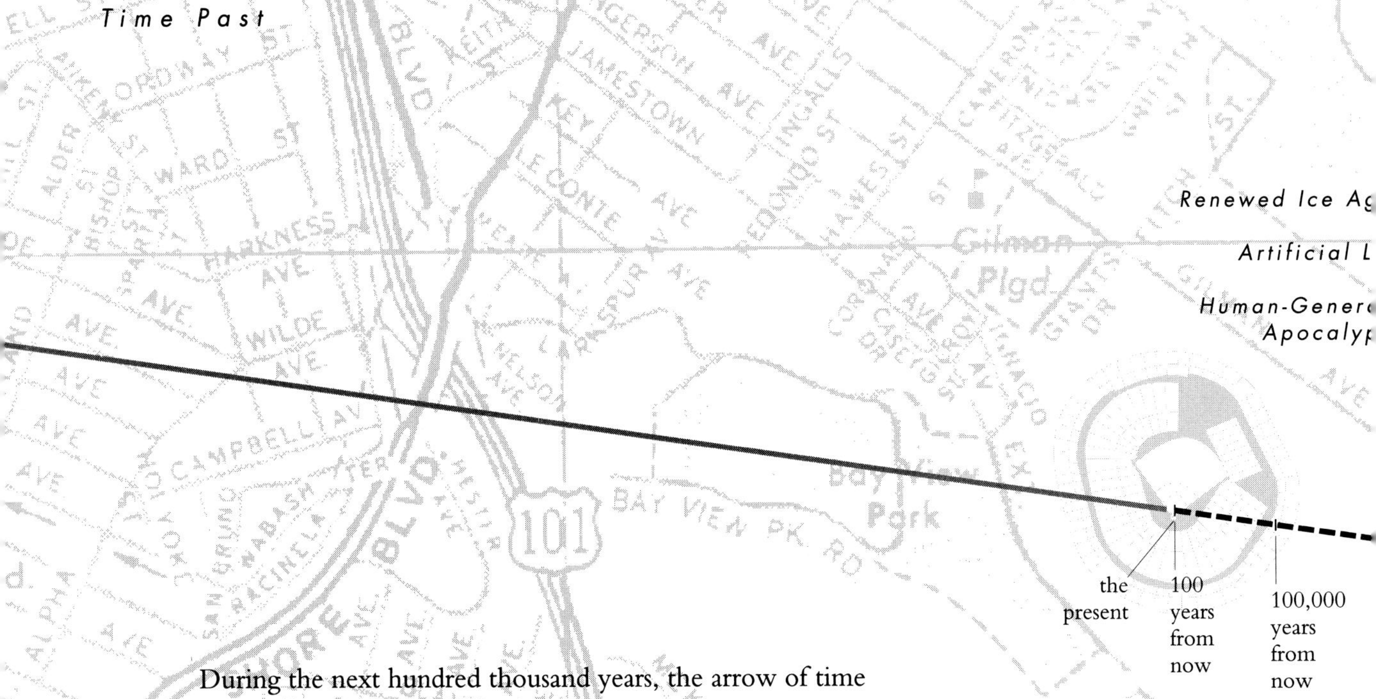

Time Past

Renewed Ice Ag

Artificial L

Human-Gener
Apocaly

the
present

100
years
from
now

100,000
years
from
now

During the next hundred thousand years, the arrow of time
moves across the other side of Candlestick Park. Within half a
million years it reaches San Francisco Bay.

ONE MILE = ONE MILLION YEARS

100 years from now to 1 million year from now

South
Basin

ficial Life Dominant?

Colonization of Other Star Systems?

Cosmic Impact?

Candlestick Point
State Rec. Area
(Under Development)

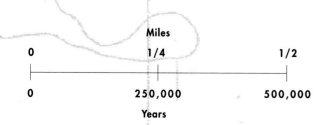

500,000 years
from now

1 million years from now

Miles

0	1/4	1/2

0	250,000	500,000

Years

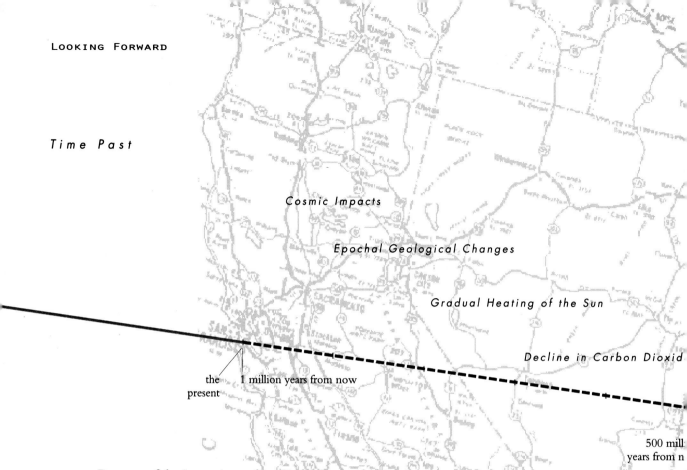

Time Past

Cosmic Impacts

Epochal Geological Changes

Gradual Heating of the Sun

Decline in Carbon Dioxid

the present

1 million years from now

500 mill
years from n

Because of the increasing radiance of the sun, if for no other reason, our species is unlikely to survive the vast sweep of time represented here. If we are to survive, it will almost certainly be by taking to space. If that occurs and humanity disperses, new species are likely to evolve from the human root.

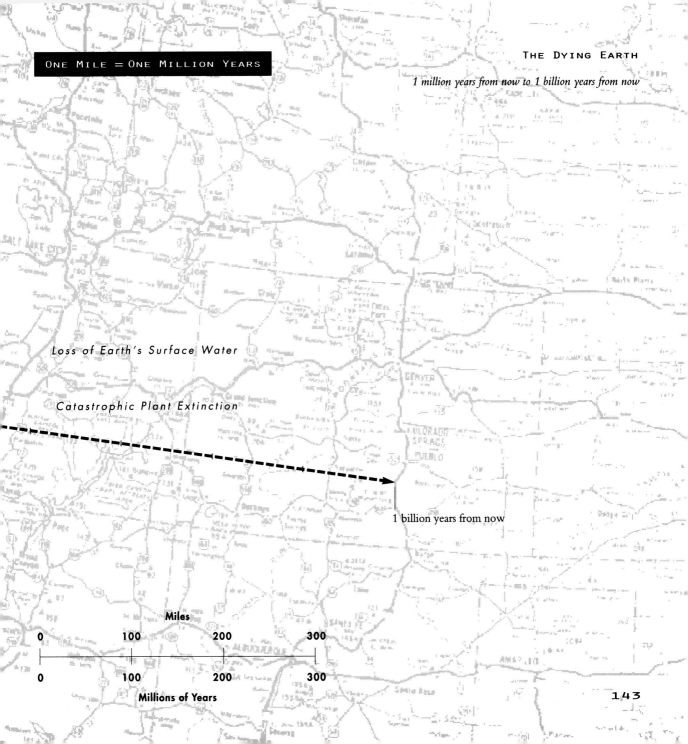

ONE MILE = ONE MILLION YEARS

THE DYING EARTH

1 million years from now to 1 billion years from now

Loss of Earth's Surface Water

Catastrophic Plant Extinction

1 billion years from now

Miles

| 0 | 100 | 200 | 300 |

| 0 | 100 | 200 | 300 |

Millions of Years

temperature exceeds tolerance for most microbes

average global temperature exceeds 50°C (122°F)

only bacteria and protists survive

sun expands into red giant phase, consuming the Earth and ending its existence

last of Earth's water eliminated; termination of all remaining life

sun condenses into a superdense white dwarf star

the present

1 billion years from now

5 billion years from now

The arrow of time moves on. The increasing heat of the sun eliminates water and terminates remaining life on Earth. Thereafter the sun's active cycle culminates in a gigantic expansion of glowing hot gas that incinerates and possibly envelops our planet. Eventually the sun condenses into a superdense white "dwarf," which shines with diminishing brilliance until it burns out.

1 billion years from now to 15 billion years from now

ONE MILE = ONE MILLION YEARS

10 billion years from now

15 billion years from now

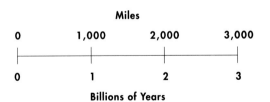

Miles

0	1,000	2,000	3,000

0	1	2	3

Billions of Years

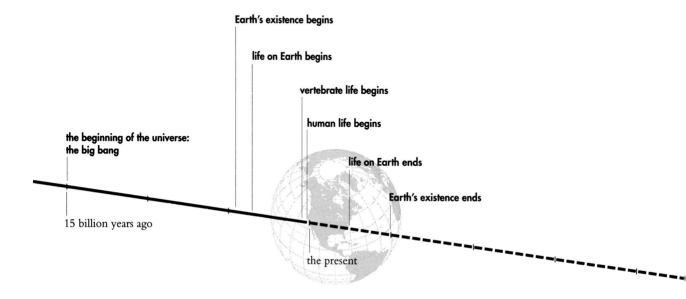

Earth's existence begins

life on Earth begins

vertebrate life begins

human life begins

life on Earth ends

Earth's existence ends

**the beginning of the universe:
the big bang**

15 billion years ago

the present

Science is rich with theories but lacking certainty as to whether the universe is "open" and will continue expanding forever, is "flat" and will reach a state of equilibrium, or is "closed" and will eventually cease to expand and then contract at an accelerating rate, ending in a "big crunch." This visualization is of a closed scenario. If, as is perhaps more likely, the universe keeps expanding forever, the arrow of time will continue forever, too.

universe stops expanding and begins contracting

the end of time:
the big crunch

70 billion years from now

Miles

0   5,000

0   5   10

Billions of Years

APPENDIX

Note: sizes and distances are approximate

Sizes of Objects Used as Reference
A grain of sand is a hundreth of an inch in diameter.
A baseball is 3 inches in diameter.
A baseball park is 1,320 feet (a quarter of a mile) across.
The Earth is 8,000 miles in diameter.

Sizes and Distances of Representative Celestial Objects
The moon is 240,000 miles away.
The moon is 2,000 miles in diameter.
The sun is 93 million miles away.
The sun is 860,000 miles in diameter.
The solar system is 7 billion miles across.
The nearest star is 26 trillion miles (4.33 light-years) away.
Our galaxy is 100,000 light-years across.
The distance to the end of the observable universe is 15 billion light-years?

Scales Commonly Used by Astronomers
93 million miles (the distance from the Earth to the sun) = one astronomical unit (1 A.U.)
5.88 trillion miles = one light-year
3.26 light-years = one parsec
3.26 million light-years = one megaparsec

Note: sizes are approximate; comparisons are linear

Comparisons

A baseball is 300 times as far across as a grain of sand.

A baseball park is 5,280 times as far across as a baseball.

The Earth is 32,000 times as far across as a baseball park.

The sun is 110 times as far across as the Earth.

The solar system is 8,300 times as far across as the sun.

The Milky Way galaxy is 85,000,000 times as far across as the solar system.

The distance to the end of the observable universe is 300,000 times the distance across the Milky Way galaxy?

Note: sizes are approximate and may vary greatly among individuals and species.

Sizes of Representative Microscopic and Near-Microscopic Objects

cat flea: 2 millimeters long

stentor: 800 micrometers long

rotifer: 400 micrometers long

pinnularia: 200 micrometers long

peridinium: 70 micrometers across

pollen grain: 16 micrometers across

red and white blood cells: 8 micrometers across

E. coli bacterium: 2 micrometers long

human chromosome: 250 nanometers wide

T4 virus: 220 nanometers long

earthworm hemoglobin molecule: 26 nanometers across

water molecule: 300 picometers across

gold atom: 250 picometers across

proton: 2 fentometers across

Metric Scale and Equivalencies

one mile = 1.6 kilometers (1600 meters)

one foot = .304 meters; 30.4 centimeters; 304 millimeters

one inch = 2.5 centimeters; 25.5 millimeters

one hundredth of an inch = .25 millimeters; 250 micrometers

1,000 micrometers = 1 millimeter

1,000 nanometers = 1 micrometer

1,000 picometers = 1 nanometer

1,000 fentometers = 1 picometer

Notes:

Micrometers are sometimes referred to as "microns."

Scientists sometimes find it convenient to describe sizes
in terms of Ångström units:

1 Ångström unit = 100 picometers

10 Ångström units = 1 nanometer

100 Ångström units = 10 nanometers

British Usage

A "thousand million" is equivalent to an American "billion."

A "billion" is equivalent to an American "trillion."

A cricket ball is almost precisely the size of a baseball.

Acknowledgments

The following material was reproduced in this book by arrangement with the respective owners:

Cover Photomontage:
(also reproduced on pp. 7 and 75)
Digital Photomontage: ©1994 Wells Packard
Photo of Candlestick Park: © 1986 Robert W. Cameron and Company, Inc.
Earth photo: NASA.

Maps and Globe of Earth:
San Francisco Bay area: from the Rand McNally Road Atlas © 1994 Rand McNally,
    R.L 94-S-71: pp. 9, 25, 57, 87, 103, 127.
From the Gousha maps: used by permission of the publisher, H.M.Gousha, a Division of
    Simon & Schuster, New York:
City of San Francisco © 1991 H.M.Gousha, Inc.: pp. 21, 22, 23, 43, 55, 88, 89, 104, 105,
    129, 130, 131, 140, 141;
North America: Gousha Interstate Road Atlas © 1993 H.M.Gousha, Inc.: pp. 11, 13, 33,
    84, 85, 101;
Central and Western United States: © 1992 H.M.Gousha, Inc.: pp. 31, 59, 125, 142, 143.
Globe of Earth: Cartesia Software: pp. 15, 35, 45, 51, 61, 99, 121, 123, 144, 146.
Candlestick Park diagram: courtesy of the San Francisco Giants

Images:
PART I — LOOKING OUTWARD
Galaxies in Leo: Mt. Palomar telescope, California Institute of Technology: p. 1.
Globular cluster: ultraviolet imaging telescope, NASA: pp. 2 and 3.
Great Galaxy in Andromeda: Mt. Wilson telescope, California Institute of Technology:
    p. 59.
Galaxy in Ursa Major; NGC 5457: Mount Palomar telescope, California Institute of
    Technology: pp. 47, 57, 59.
Large-scale galaxy structures: computer artist's conception adapted and extrapolated from
    galactic surveys by Margaret J. Geller, John P. Huchra, and Luis da Costa, Harvard-
    Smithsonian Center for Astrophysics/original graphics by Emilio Falco: pp. 62–69.

PART II — LOOKING INWARD

Note: "PR" signifies Photo Researchers, Inc.

Collection of algae: M. I. Walker/PR: pp. 70, 71.

Grasshopper: James Burgess: pp. 76, 96.

Ant: James Burgess: pp. 77, 79, 85, 97.

No-see-um: N. Y. S. Dept. of Agriculture: pp. 77, 85, 97.

Cat flea: Biophoto Associates/PR: pp. 77, 78, 85, 97, 98.

Millipede: Biophoto Associates/ PR: pp. 77, 79, 85, 97, 99.

Grain of sand: SPL/PR: pp. 77, 79, 80, 85, 87, 96, 99, 100.

Water bear: SPL/PR: pp. 77, 79, 81, 85, 97, 99, 101.

Rotifer: Walter Dawn/PR: pp. 77, 79, 85, 93, 97, 99.

Stentor: Eric Grave/PR: pp. 77, 79, 85, 96, 97, 99.

Micrasterias: courtesy of Dr. Gary Floyd: pp. 99, 101.

Pinnularia: courtesy of Dr. Larry R. Hoffman: pp. 79, 81, 87, 99, 101.

Peridinium: courtesy of Dr. Gary Floyd: pp. 81, 87, 99, 101.

Nutmeg pollen: courtesy of Dr. Gil Brenner: pp. 81, 87.

Amoeba: PR: pp. 79, 85, 87, 97, 99, 101.

Red blood cells: SPL/PR: pp. 81, 87, 101.

White blood cell: SPL/PR: pp. 81, 87, 89, 101, 103.

Ovum and sperm: David M. Phillips/PR: pp. 81, 87, 101.

E. coli bacterium: G. Morti/SPL/PR: pp. 89, 91, 103, 105.

Human chromosome: CNRI/PR: pp. 89, 105.

Hemoglobin molecule: SPL/PR: pp. 91, 92, 93, 106, 107.

Ganglion (thorn) cell: courtesy of Dr. R. W. Rodieck: pp. 80, 86, 100.

Mitochondrion from primate retina: courtesy of Dr. Robert Smith; also thanks to Dr. Y. Tsukamoto, P. Masarachia, and Dr. P. Sterling for providing the electron micrograph: pp. 89, 103.

Adenovirus: James Burgess: pp. 91, 105.

Streptococci bacterium: Richard Brightfield: p. 103.

Lyme spirochete bacterium: courtesy of Dr. Russell C. Johnson; © Yale Journal of Biology and Medicine: pp. 89, 102, 103.

Epf1 virus: courtesy of Dr. Loren Day and Dr. Leon Kostrikis: pp. 91, 105.

T4 virus; tobacco rod virus: electron micrographs taken on the STEM at Brookhaven National Laboratory; courtesy of Dr. Martha Simon: pp. 91, 104, 105.

Sodium chloride crystal; polyethylene chain: courtesy of Dr. Dale J. Meier: pp. 93, 107, 109.

Water molecule: Richard Brightfield: pp. 93, 107, 109.

Gold atom: Dr. Mitsuo Ohtsuki/PR pp. 109, 114.

PART III — LOOKING BACKWARD AND LOOKING FORWARD

Ocean: courtesy Bethany Johns: pp. 118, 119.

Most of the following individuals did not see even portions of the text and illustrations, nor was it practicable to follow all the many suggestions made, so the author alone is responsible for any inaccuracies and shortcomings of this book. Nonetheless, special thanks to the following, and to others not here named, for their advice and information: James Burgess, Helena Curtis, Dr. Loren Day, Joan Deutsch, Dr. Seth Digel, Dr. Gary L. Floyd, Dr. William A. Gutsch, Jr., Dr. Htun Han, Dr. Frank N. Jones, Dr. Leon Kostrikis, Dr. Larry Liddle, Dr. J. David Nightingale, Dr. Martha Simon, Dr. Robert Smith, William Speare, Dr. John R. Taylor, and Dr. Charles Whitney.